Heat Transfer - Advances in Fundamentals and Applications

Edited by Hafiz Muhammad Ali and T. M. Indra Mahlia

Published in London, United Kingdom

Heat Transfer - Advances in Fundamentals and Applications
http://dx.doi.org/10.5772/intechopen.107587
Edited by Hafiz Muhammad Ali and T. M. Indra Mahlia

Contributors
Sharat V. Chandrasekhar, Poodipeddi V Suryanarayana, Udaya B. Sathuvalli, Shamoon Jamshed, Tatsuhiko Aizawa, Hiroki Nakata, Takeshi Nasu, Marcello Garavaglia, Marco Rottoli, Marta Mantegazza, Fabio Grisoni, Nikolai Kobasko, Baljit Bhathal Singh, Jorge Sergio Téllez-Martínez, Hugo Enrique Alva-Medrano, Miriam Zulma Sánchez-Hernández, María Janeth Vega-Flores, Abel Alberto Pintor-Estrada, Nicolás David Herrera-Sandoval, Montse Vilarrubí

First published in London, United Kingdom, 2024 by IntechOpen
IntechOpen is the global imprint of INTECHOPEN LIMITED, registered in England and Wales, registration number: 11086078, 5 Princes Gate Court, London, SW7 2QJ, United Kingdom

British Library Cataloguing-in-Publication Data
A catalogue record for this book is available from the British Library

Additional hard and PDF copies can be obtained from orders@intechopen.com

Heat Transfer - Advances in Fundamentals and Applications
Edited by Hafiz Muhammad Ali and T. M. Indra Mahlia
p. cm.
Print ISBN 978-1-80356-639-9
Online ISBN 978-1-80356-640-5
eBook (PDF) ISBN 978-1-80356-674-0

For EU product safety concerns:
IN TECH d.o.o., Prolaz Marije Krucifikse Kozulić 3, 51000 Rijeka, Croatia,
info@intechopen.com or visit our website at intechopen.com.

Meet the editors

Hafiz Muhammad Ali is an Associate Professor of Mechanical Engineering and an affiliate of the Interdisciplinary Research Center for Renewable Energy and Power Systems (IRC-REPS) at King Fahd University of Petroleum and Minerals, Saudi Arabia. He was recognized as a highly cited researcher in engineering in 2021 and 2022. He is a renowned scientist in thermal engineering and an active editorial member of several high-ranking international journals, including the *Journal of Thermal Analysis and Calorimetry*, *Heat Transfer*, and *Applied Nanoscience*.

T. M. Indra Mahlia, an Australian engineer, is a highly cited researcher and distinguished professor at the School of Civil and Environmental Engineering, University of Technology Sydney (UTS), Australia. A cornerstone of UTS's Centre for Water and Wastewater Technology, Dr. Mahlia's expansive research focuses on energy, techno-economic analysis, and circular economy. He has secured more than $5.5 million in project funding. *The Australian* spotlighted his sustainable energy leadership in 2019. Since 2018, Dr. Mahlia has emphasized mentoring, with two of his former Ph.D. students and two postdoctoral fellows becoming highly cited researchers.

Contents

Preface

This new book, *Heat Transfer – Advances in Fundamentals and Applications*, is unique in its coverage of advanced features and applications of heat transfer. It contains seven chapters organized into two sections.

Section 1 includes three chapters that cover the fundamental advances in heat transfer. Chapter 1, "Analysis of a Fundamental Procedure for Solving the Inverse Heat Conduction Problem Applied to Simple Systems", considers various simple systems and different methods to explore the solution of inverse heat conduction. Chapter 2, "Heat Transfer Fundamentals Concerning Quenching Materials in Cold Fluids", explores and discusses heat transfer's basic aspects while quenching materials within a fluidic environment. Chapter 3, "A Brief Review of Techniques of Thermal Enhancement in Tubes", provides insight into the heat transfer enhancement methods used for internal flows, specifically focusing on tubes.

Section 2 includes five chapters that explore advances in heat transfer applications. Chapter 4, "Advanced Shell-and-Tube Longitudinal Flow Technology for Improved Performances in the Process Industry", discusses the advanced applied technological improvements in shell-and-tube heat exchangers for various industry applications. Chapter 5, "Heat Transfer Mechanisms in Petroleum and Geothermal Wellbores", focuses on the applied nature of heat transfer mechanisms during wellbores in petroleum and geothermal applications. Chapter 6, "Heat Transportation by Acicular Micro-Textured Device with Semi-Regular Alignment", discusses applied problems. Chapter 7, "Analysis and Optimization of Heat Transport for the Purpose of Maximizing the Potential of Solar Ponds in Sustainable Energy Applications", examines solar ponds, emphasizing their sustainability. Finally, Chapter 8, Perspective Chapter: Smart Liquid Cooling Solutions for Advanced Microelectronic Systems".

We hope that this book will be helpful for readers working in the fundamental and applied domains of heat transfer who wish to update their knowledge.

Hafiz Muhammad Ali
Department of Mechanical Engineering,
Interdisciplinary Research Center for Renewable Energy and Power Systems (IRC-REPS),
King Fahd University of Petroleum and Minerals,
Dhahran, Saudi Arabia

T. M. Indra Mahlia
School of Civil and Environmental Engineering,
University of Technology Sydney,
Sydney, Australia

Section 1

Heat Transfer – Advanced Fundamentals

Chapter 1

Analysis of a Fundamental Procedure for Solving the Inverse Heat Conduction Problem Applied to Simple Systems

Jorge Sergio Téllez-Martínez,
Miriam Zulma Sánchez-Hernández, María Janeth Vega-Flores,
Abel Alberto Pintor-Estrada, Hugo Enrique Alva-Medrano and
Nicolás David Herrera-Sandoval

Abstract

Heat transfer phenomena develop in various natural and artificially created processes. Fundamental laws of physics allow the transfer mechanisms to be classified; however, describing the phenomena is relatively complex, even if the analysis is limited to conduction. In particular, to determine the temperature distribution in a solid body, the definition of the boundary conditions that perturb it is required. Such conditions mathematically model a fluid's hydrodynamic and thermodynamic behavior, and when the temperature differences are significantly high, the flow by radiation. It is then complex to define the functions of the thermal boundary conditions and solve the well-posed problem. Naturally, nonlinear system results and applying numerical methods are constant in the analysis. However, a unique solution for the thermal field in a solid does ensure. Alternatively, the scheme of the discretization of the system allows us to propose that through the knowledge of a fraction of the thermal field, the boundary condition is quantified independently of its nature. Such a procedure is called inverse analysis and has the characteristic of not satisfying the single solution criterion. However, some cases of interest can treat, and the estimate is guaranteed to be highly accurate.

Keywords: transport phenomena, IHCP, heat and mass flux, estimated boundary conditions, sequential function specification

1. Introduction

The researchers Dowding and Beck [1] define that inverse heat conduction problems (IHCPs) can be categorized in various ways, (1) by the solution technique or algorithm (Function Specification, Regularization, Laplace Transform, Conjugate

Gradient, Mollification, etc.), (2) by the solution method (Duhamel's Theorem, Finite Difference Method, Finite Element Method, etc.), and (3) by the time domain (Stoltz Method, Sequential, and Complete Domain Method). Beck et al. [2] developed a whole theory, and other researchers have used it to adapt it to particular objectives [3–5]. Therefore, the characteristic of the inverse problems of not meeting the criteria of existence and uniqueness of the solution has promoted the implementation of more robust mathematical methods in each classification of the IHCP to optimize its stability since minor errors in the measured data can induce significant inaccuracies in the estimated variable [6–12]. The analyses have oriented the study with an analytical mathematical approach [13–15] and to applying numerical solutions in various physical phenomena, particularly thermal phenomena [16–19].

2. Inverse heat conduction problem

Transient thermal phenomena are relatively complex to be described through mathematical formulations. However, it is possible to obtain acceptable approximations in systems that can be bounded. In this regard, the problems posed by Beck et al. [2] for a solid with cylindrical geometry will be formulated and developed. With this objective, the numerical method based on heat flow balances results in sufficient precision and easy programming.

It begins with defining the heat conduction equation in its differential form for the cylindrical coordinate system represented by the variables r, θ, and z. Eq. (1) specifies the material's properties through k, ρ, and Cp; thermal conductivity; density; and thermal capacity, respectively. The field or dependent variable T is temperature, and t represents the variable time. Finally, $\dot{q}$ defines the heat generated or consumed per unit volume.

$$-\left[\frac{1}{r}\frac{\partial}{\partial r}\left(-kr\frac{\partial T}{\partial r}\right)+\frac{1}{r}\frac{\partial}{\partial \theta}\left(-k\frac{1}{r}\frac{\partial T}{\partial \theta}\right)+\frac{\partial}{\partial z}\left(-k\frac{\partial T}{\partial z}\right)\right]+\dot{q}=\frac{\partial \rho C_p T}{\partial t} \tag{1}$$

Mainly, to clarify the process of estimating thermal boundary conditions through inverse analysis, interference from internal heat "sources" or "wells" is not desired, so the $\dot{q}$ term did cancel. From this new form of Eq. (1), the formulations of the specific problems will be developed, which will address the inverse analysis.

2.1 One-dimensional analysis in the radial direction of heat transfer—An *ill-posed* problem for estimating the function in the active boundary

Considering the dimensional restriction to a flow direction simplifies the problem. Additionally, although the physical properties associated with a material depend on temperature, due to the sequential numerical solution strategy that will be adopted, constants can be defined at the instant of calculation time. That is, it is evident that the temperature distribution is unknown at the moment of calculation, but is know at the last instant. Therefore, if it does consider that there will be no significant variations in the properties due to relatively small temperature changes, then they are estimated and projected to the calculation time step. In this way, the well-posed or direct heat conduction problem (DHCP) is formulated according to the information in **Figure 1**.

The problem is mathematically nonlinear; therefore, numerical analysis is required to obtain a solution for the temperature variable. The finite difference method was selected based on the systems' geometric characteristics and the developments' intelligibility.

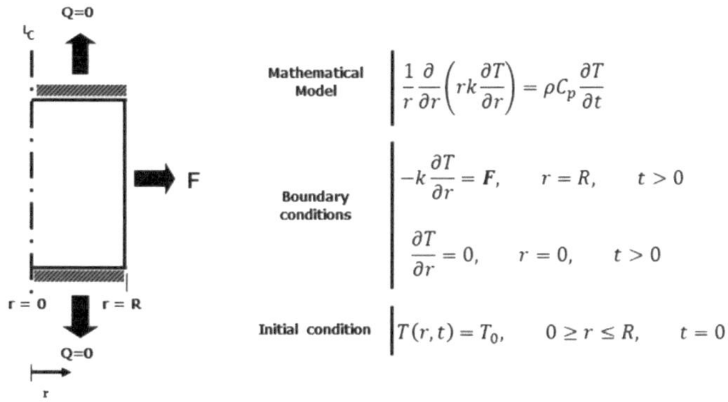

Figure 1.
The mathematical formulation of the DHCP in the radial direction and transitory state.

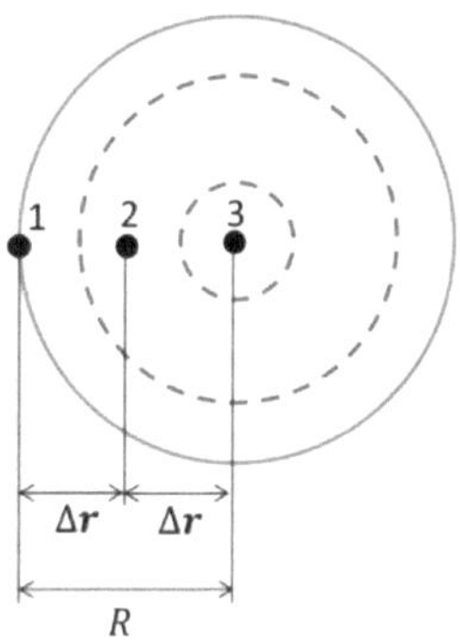

Figure 2.
Discretization of the cylindrical system considering a one-dimensional analysis in the radial direction.

In the methodology, the system domain is subdivided into adjacent sections creating so-called control volumes (CVs). **Figure 2** shows a simplified scheme to indicate the representative basic rings according to the definition of the geometry and direction of heat flow. As can be deduced, the notation Δr represents a fraction of the radius of the cylinder.

Equation (2), called the general equation of heat flux balance Q, is equivalent to the mathematical model presented in **Figure 1** and will be applied to each CV in the scheme.

$$Q_{In} - Q_{Out} = Q_{Accumulation} \tag{2}$$

By definition of Fourier's law for the heat flow entering Q_{In} and leaving Q_{Out}, in addition to the heat flow per unit volume that accumulates in a transitory state $Q_{\text{Accumulation}}$, the system of equations corresponds to the number of CV of meshed space. If a Crank–Nicolson technique approach does consider calculating the field variable concerning the transient state, the expressions in **Table 1** are obtained.

In each system considered, a mesh sensitivity analysis should do perform. It can deduce that obtaining results with greater precision of the temperature distribution requires defining a more significant number of CVs. Therefore, it is required to determine the minimum number of CVs in which no differences are detected in the calculation result. On the other hand, the parameter Δr can be heterogeneous when

VC	General heat balance equation	i
1	$q_i^t A_{In,1} - \left(-\overline{k_{i\to i+1}} A_{Out,i} \frac{T_{i+1}^t - T_i^t}{\Delta r}\right) = \rho_i C_{pi} V_1 \frac{T_i^{t+\Delta t} - T_i^t}{\Delta t}$	1
2	$-\overline{k_{i-1\to i}} A_{In,i} \frac{T_i^t - T_{i-1}^t}{\Delta r} - \left(-\overline{k_{i\to i+1}} A_{Out,i} \frac{T_{i+1}^t - T_i^t}{\Delta r}\right) = \rho_i C_{pi} V_i \frac{T_i^{t+\Delta t} - T_i^t}{\Delta t}$	$2, \dots, n-1$
3	$-\overline{k_{n-1\to n}} A_{In,i} \frac{T_n^t - T_{n-1}^t}{\Delta r} - 0 = \rho_n C_{pn} V_n \frac{T_n^{t+\Delta t} - T_n^t}{\Delta t}$	n

Table 1.
Summary of the development of the general equation of balance of heat flow Eq. (2) on the CVs in the discretization of the cylindrical system in the radial direction.

system sections are defined with a different degree of mesh. The strategy helps to focus sensitivity on important sections and to make computational time efficient.

The function **F** specified in **Figure 1** can be defined as Dirichlet, Neumann, or Robinson type. However, the general heat flux balance equation determines a Neumann-type function. Therefore, the general form of the equations can be stated as follows:

$$d_i T_{i-1}^t + e_i T_i^t + f_i T_{i+1}^t = T_i^{t+\Delta t} - g_i, i = 1, 2, \dots, n \tag{3}$$

Where d, e, f, and g represent the respective terms product of the algebraic development of the corresponding equations in **Table 1**. In this way, the system of equations of the type does generate:

$$\begin{bmatrix} e_1 & f_1 & \cdots & 0 & 0 \\ d_2 & e_2 & f_2 & 0 & 0 \\ \vdots & \ddots & \ddots & \ddots & \vdots \\ 0 & 0 & d_{n-1} & e_{n-1} & f_{n-1} \\ 0 & 0 & \cdots & d_n & e_n \end{bmatrix} \begin{bmatrix} T_1^t \\ T_2^t \\ \vdots \\ T_{n-1}^t \\ T_n^t \end{bmatrix} = \begin{bmatrix} T_1^{t+\Delta t} - g_1 \\ T_2^{t+\Delta t} - g_2 \\ \vdots \\ T_{n-1}^{t+\Delta t} - g_{n-1} \\ T_n^{t+\Delta t} - g_n \end{bmatrix} \tag{4}$$

Implementing solution strategies for matrix systems will depend on their structure, but the Thomas method, successive over-relaxation, and LU decomposition do suggest. A solution to the well-posed or direct problem is relevant because it is required to solve the *ill-posed* or inverse problem. The algorithm for solving the direct heat conduction problem establishes the core of the solution of the inverse analysis since it is used repeatedly in each cycle of estimating the thermal boundary conditions depending on the value of the stabilization parameter.

In turn, the wrongly stated or inverse problem defines that the boundary condition at $r = R$ is unknown; however, if there is knowledge of the thermal history obtained during a monitoring event at least one point within the solid. The formulation is summarized in **Figure 3**.

The implementation of Beck's theory, based on the sequential method of specification of the function, refers to the determination in each calculation step of heat flux density $q(t)$. Estimating the value of data of the function depends exclusively on the internal gradients in the solid and not on the potential extraction or supply of heat by an external medium (as foreseen, a Neumann-type boundary condition does define as an unknown, the reason over the Dirichlet and Robinson types is the natural condition of their specification through the heat flux general balance equation).

It is more common and with greater dominance that in the field of applicability of boundary conditions in heat transfer problems, Robinson-type conditions identify

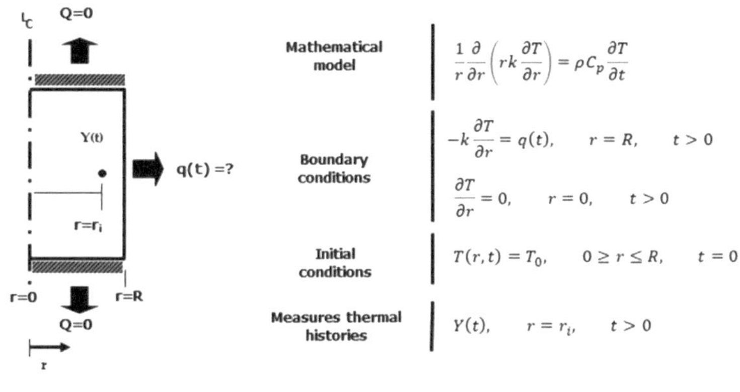

Figure 3.
Formulation of the ill-posed *problem or IHCP for the same system referenced in* ***Figure 1***.

where a heat transfer coefficient "h" explicitly expresses. The magnitude of the values of a function h represents complex behavior conditions of media whose primary heat transfer mechanism is convection. However, some data may be associated with a combination with a radiation mechanism. In addition, they are related to the temperature of the surface of the solid and to the temperature of the fluid sine through Newton's law of cooling, defined by Eq. (5).

$$q = h(T_s - T_\infty) \tag{5}$$

Particularly, it is possible to determine a function of the Robinson-type boundary condition with the solution methodology of the IHCP. However, Newton's law of cooling approach determines that once the difference between the surface and bulk temperatures of the fluid tends to zero, the function does not adequately satisfy the thermal boundary condition that causes small gradients in the solid. This condition exposes a field of research of great interest. The solution by inverse analysis could support the modification of Newton's cooling law, being used as a reference to model a mathematical complement that allows determining the appropriate value of the heat transfer coefficient at surface temperatures close to the temperature of the fluid bulk.

Returning to the specification of a Neumann-type condition is justified to describe the complete transitory heat transfer process. For this purpose, applying the norm of least squares defined by Eq. (6) as an objective function is proposed.

$$S = \sum_{l=1}^{\ell} \sum_{j=1}^{n} \left(Y_{j,M+l-1} - T_{j,M+l-1}\right)^2 \tag{6}$$

The objective function establishes that the square of the difference between the registered and calculated temperatures, proposing a value of the border condition, which will be minimized. Since more recorded histories can integrate, then the minimization of the sum of the squares of all the differences is considered. In addition, since there is the possibility of knowing the temperature data after a reference instant, as much data as necessary do use to obtain weights of the immediate "future" and is also considered in the minimization. This last definition does use to stabilize the determination of the magnitude of the value of the boundary condition. As it is associated with the sequence of calculation time steps, the amount of data considered is called the number of future time steps (ℓ). Naturally, to achieve stabilization, the minimum value of the number must be two, and for each calculation, the matching value of the proposed boundary condition does consider ($q_{M-1} = 0$ is recommended). Following the above, the mathematical development for determining the boundary condition's value does obtain by the expression of Eq. (7).

$$q_M - q_{M-1} = \frac{1}{\Delta M}\sum_{l=1}^{\ell}\sum_{j=1}^{m}\left(Y_{j,M+l-1} - T^*_{j,M+l-1}\right)\left(T^*_{j,M+l-1;q}\right),$$
$$\Delta M = \sum_{l=1}^{l}\sum_{j=1}^{m}\left(T^*_{j,M+l-1;q}\right)^2, \tag{7}$$
$$T^*_{j,M+l-1;q} = \frac{\partial T_{j,M+l-1}}{\partial q_M}$$

As can be deduced, considering $\ell = 2$ or greater implies solving the same number of times the formulation of the DHCP. In addition, as seen in Eq. (7), according to the term called sensitivity coefficient, it is necessary to solve again the same number of times the result of the derivation of the same formulation. **Figure 4** shows the compendium of the corresponding development.

Since a large amount of information must be processed to specify a single piece of data for the function $q(t)$, it is convenient to focus efforts on developing a computer application. **Figure 5** contains the flowchart of the pre-explanation calculation algorithm to obtain the sequential specification of the complete transient event function.

$$S_Y = \left[\frac{N}{N-1}\sum_{\aleph=1}^{N}\left(Y_\aleph(t) - \hat{Y}_\aleph(t)\right)^2\right]^{\frac{1}{2}} \tag{8}$$

Currently, there is the advantage of having efficient computer systems. In this sense, the algorithm can modify to include decision-making parameters to optimize the solution's stability without influencing the estimation's precision. For example, a weighted error parameter at each calculation step is defined by Eq. (8). The equation represents the calculation of the standard deviation by relating the recorded and calculated thermal histories to the function of the estimated boundary condition.

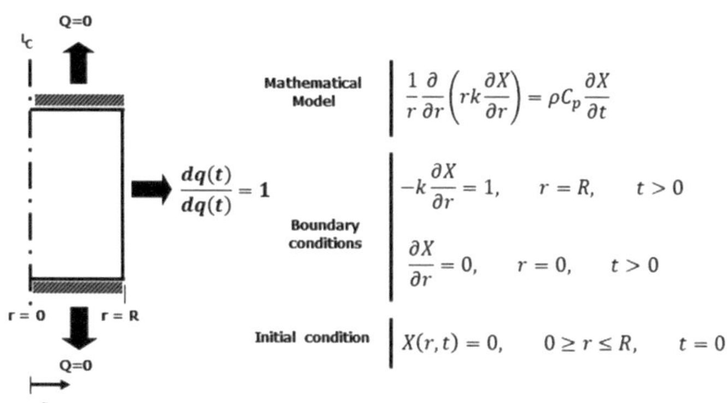

Figure 4.
Mathematical formulation for the calculation of the sensitivity coefficients for the solution of the IHCP of the cylindrical system with radial unidirectional flow and a single boundary condition.

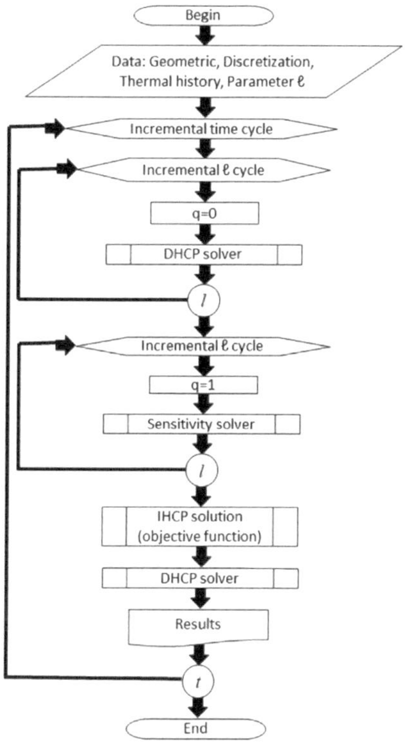

Figure 5.
Flowchart of the algorithm for solving the IHCP.

The degree of error will be associated with the "quality" of the information of each thermal history fed to the calculation algorithm and the designation of the number of future time steps (l). A high value of l helps to stabilize the IHCP solution when the quality of the thermal history is low; that is, the imprecision due to the effect of noise introduced in the data does minimize. However, when the quality of the thermal history is high, the magnitude of the function $q(t)$ values is underestimated. The underestimation can be noted graphically by the trend of a smooth transition of the function $q(t)$ curve, above all, in moments where the temperature could change suddenly. In this way, with a previous analysis of the quality of the thermal history, the error parameter can be used to include an adjustment process of l that does not imply a critical underestimation of the function $q(t)$ concerning the thermal information used.

Fundamentally, the previous implies that before analyzing natural systems, exercises are carried out to verify the formulations and the coding of algorithms through submission to benchmark cases. The fundamental methodology fulfills at least the following steps:

- Establish a system with a known thermal boundary condition and calculate thermal histories solving the well-posed or direct heat conduction problem. Based on a thermodynamic analysis, negative data indicates heat extraction from the solid.
- Choose at least one thermal history calculated at a specific position, preferably the location close to the border of interest.

- Provide at least one thermal history regarding the point before the application that solves the wrongly stated or inverse problem.
- Compare the estimated thermal boundary condition function, as well as the thermal history calculated with it, with the corresponding data from steps 1 and 2. A reliable IHCP solution tool obtains if there are no significant differences. Otherwise, one must try to stabilize the solution through the number of future time steps or modify the mathematical formulation.

In **Figure 6**, the graphs obtained from the previous steps are summarized. It can be noted that abrupt magnitude changes in the boundary condition represent extreme test points for the IHCP solution method. The bias presented by the estimated function establishes a research topic that should be dealt with separately from the discussion here. However, the method can project to natural systems with the results obtained.

The sensors and data acquisition systems must calibrate before a metrology certifying unit. On the other hand, the systems coupled to the measurement target must be shielded as well as possible to avoid environmental disturbances and human error. In essence, for diverse systems, the quality of the measurements of the field variable used to estimate boundary conditions represents the paradigm of inverse analysis. In addition, there is the component of access to the system of interest, often made impossible by the operating conditions, the dimensions, or the specific location. However, with the current advancement of filters in electronic and computer systems, it has not been necessary to increase the complexity of mathematical methods to stabilize the solution of *ill-posed* problems.

An example of a direct, unfiltered measurement of thermal history on a cylindrical specimen is presented in **Figure 7**. The trend from the beginning (t = 0 s) to end (t = 34 s) in the measurement represents a decrease in temperature. Therefore, it is defined as a cooling curve. The measurement was obtained in a cylinder that meets unidirectional flow conditions by keeping an equivalent Height-Diameter ratio of 4:1. The AISI 304-type stainless steel material guarantees that there will be no exothermic or endothermic internal reactions per unit volume equivalent to the $\dot{q}$ term. Installing a type K bayonet thermocouple for temperature sensing required drilling with a 1 mm

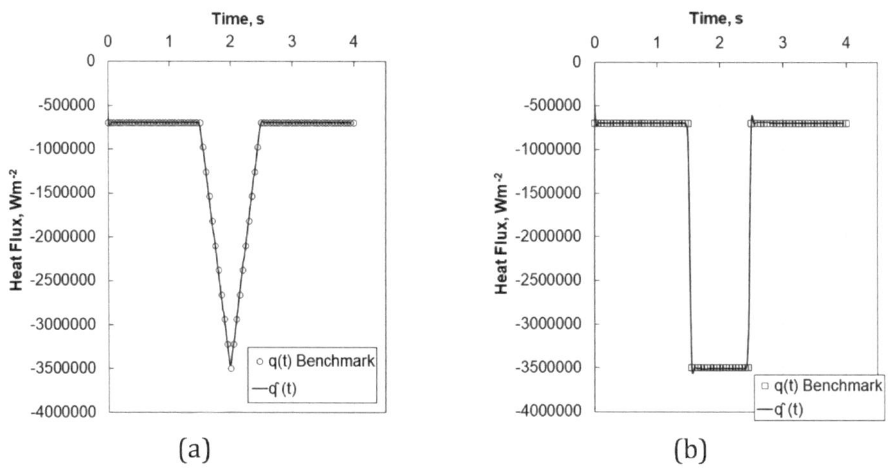

Figure 6.
Benchmark boundary conditions q(t) *with (a) triangular and (b) pulse shapes, used to assess the accuracy of the estimate* $\hat{q}(t)$ *by solving the IHCP.*

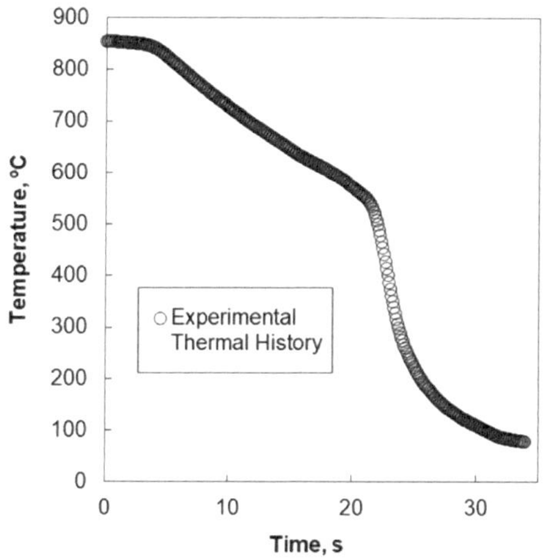

Figure 7.
Thermal history recorded inside a cylindrical specimen where heat flow is obtained by cooling from a high temperature. In this case, the thermal history $Y(t)$ *is called the* "cooling curve".

diameter drill bit 1.6 mm from the curved surface from the top to half the height of the specimen. The diameter of the analyzed specimen is 12.7 mm.

The results of estimating the boundary condition using the thermal history and solving the IHCP are expressed as the function $\hat{q}(t)$ in the graph of **Figure 8a**. In turn, with the history heat flux function, the temperature field in the cross-section of the cylinder was determined (solving the DHCP), and the thermal history of the CV where the thermocouple does virtually locate were identified. **Figure 8b** graph plots the measured $Y(t)$ (empty circular marker) and calculated $\hat{Y}(t)$ thermal histories (continuous line). As can be seen, no significant differences are observed between the thermal histories, indicating that the boundary condition has done the estimate with good precision.

However, when analyzing the standard deviation weighting, an optimization modification did implement in the IHCP solution algorithm, specifically, the option to

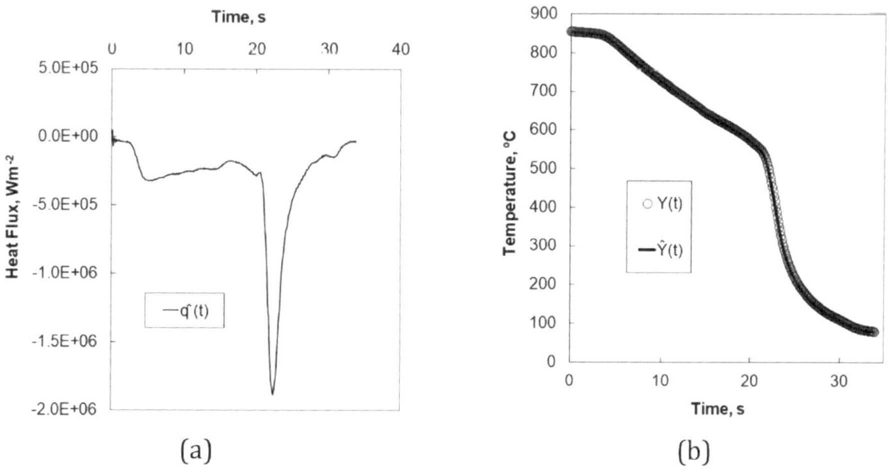

Figure 8.
Results of the application of the IHCP solution algorithm for the determination of (a) the function of the heat flux history $\hat{q}(t)$ *and (b) the thermal history in the virtual position of the thermocouple* $\hat{Y}(t)$ *(solid line).*

auto-adjust the number of future time steps between 2 and a user-defined upper limit, comparing the results to the standard deviation. In this process, the value of l is increased by one unit in each loop cycle. If the magnitude of the deviation is less than the previous one, then a new increment is performed. Otherwise, the result does keep with the last best-estimated value. **Figure 9** shows the standard deviation calculation called the residual at each calculation step through the dashed curve. When implementing the recursive process for determining ℓ considering a maximum value of 10, it can be seen through the dotted line that the residual decreased. The most notable difference does find between 23 and 30 s. The history of ℓ is also recorded and does show by the solid line with filled square symbols.

The result of the IHCP solution with ℓ adjustable and ℓ = 2 does compare in the curves of **Figure 10**. As seen in the heat flux histories curves of **Figure 10a**, a value of ℓ greater than 2 tends to underestimate. However, it also indicates a tendency toward stabilizing the solution if an alteration occurs due to disturbances in the experimental thermal history. In this case, the comparison of the deviation percentages of the thermal accounts calculated in the virtual position of the thermocouple in **Figure 10b** shows that the increase of ℓ improved the estimation of the function.

According to the above, training in understanding the phenomena of heat transfer by the conduction mechanism and the numerical analysis allows us to identify the integrity and precision of the estimation of boundary conditions by solving the problems posed in the reverse form. In addition to the materials' constitutive properties, the analysis's complexity increases with the number of active boundaries and dimensional coordinates, as seen in the studies in the following sections.

2.2 One-dimensional analysis of heat flow in radial direction: *Ill-posed* problem for estimating one function in each frontier

Modifying the system of the previous section by making a circular cut centered on the axial axis generates a new thermal boundary without insulation. Therefore, it is

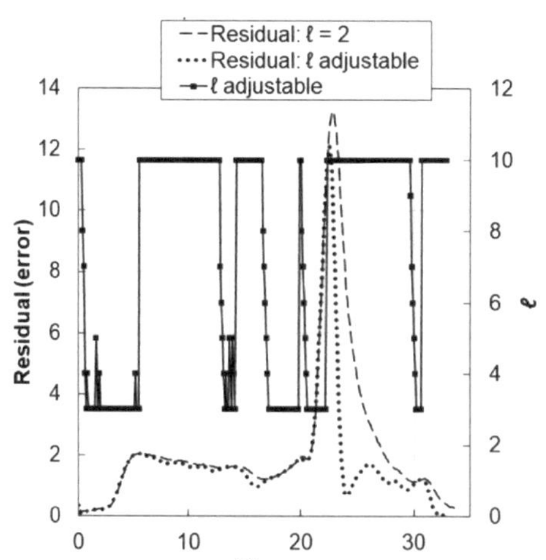

Figure 9.
Calculate the (residual) standard deviation used to determine the value of the number of future time steps ℓ (solid line with filled square marker). The curves (a) dashed line and (b) dotted line show the differences in the magnitude of the residual by using a constant and adjustable value of ℓ, respectively.

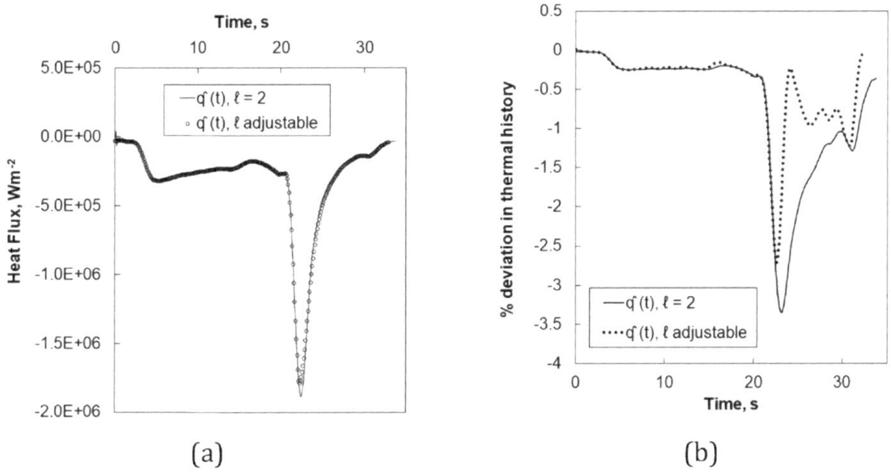

Figure 10.
Differences in (a) the estimation of the heat flux function with a constant value of ℓ = 2 (continuous line) and adjustable to a maximum of 10 (open circles markers), evidenced by the comparison of (b) the percentage of deviation between the experimental temperature history and that calculated with ℓ = 2 (solid line) and with adjustable ℓ (dotted line).

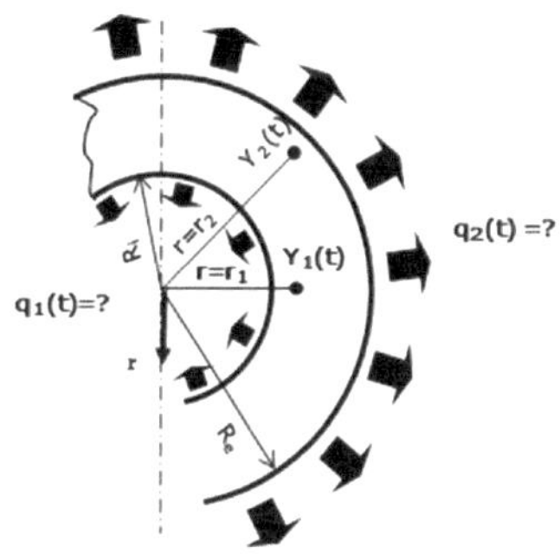

Mathematical model

$$\frac{1}{r}\frac{\partial}{\partial r}\left(rk\frac{\partial T}{\partial r}\right) = \rho C_p\frac{\partial T}{\partial t}$$

Boundary conditions

$$-k\frac{\partial T}{\partial r} = q_1(t) = ?, \quad r = R_i, \quad t > 0$$

$$-k\frac{\partial T}{\partial r} = q_2(t) = ?, \quad r = R_e, \quad t > 0$$

Initial conditions

$$T(r,t) = T_0, \quad R_i \geq r \leq R_e, \quad t = 0$$

Measures thermal histories

$$Y_j(t), \quad r = r_j, \quad t = 0, \quad j = 1,2$$

Figure 11.
Formulation of the IHCP for a cylindrical system with the unidirectional flow in the radial direction with two active boundaries.

necessary to formulate the new IHCP considering a dual estimation procedure. **Figure 11** outlines the case study with the following characteristics:

- The concentric conditions of the surfaces in the radial coordinate determine that the gradients in the angular direction of the system are not significant. It is also established that the height-diameter ratio meets the minimum requirement of 4:1.

- The heat conduction in the solid does determine by ignoring the discontinuity in the material due to the insertion of temperature sensors (instrumentation with thermocouples), and the absence of sources or internal heat sinks in the solid does consider.

- The material is considered isotropic, and the thermophysical properties (k, ρ, and Cp) are a function of temperature.

- A homogeneous heat flux history $q1(t)$ can specify the boundary condition on the inner surface.
- A homogeneous heat flux history $q2(t)$ can specify the boundary condition on the external surface.
- The initial temperature field is known.
- At least two points, information is available on the temperature distribution inside the body to adequately estimate the unknown thermal boundary conditions.

The solution of the mathematical formulation is obtained again by applying numerical methods mainly due to:

1. The nonlinearity of the partial differential equation in the transient state and thermophysical properties dependent on the temperature that governs the heat conduction problem,
2. The nonlinearity of the border conditions, as well as,
3. The vast magnitude of the time domain (depending on the density of experimental information) in which the phenomenon does analyze.

In this way, the sequential technique of estimation of the function at intervals of analysis delimited in time (fractions of the total time of the observed phenomenon) allows transforming the nonlinear problem into a "temporarily" linear one with a solution of the system for each step of time through the approach of an "objective function" and the introduction of the concept of "number of future time steps" used for the estimation calculation.

According to Beck's approach [2], (Eq. (9)), the result of a Taylor series expansion, determines the matrix operation for calculating the temperature distribution concerning the operational boundary conditions.

$$\mathbf{T} = \mathbf{T}_{\mathbf{q}=0} + \mathbf{X}\mathbf{q} \tag{9}$$

Carrying out a slight adaptation of the notation to include a positioning matrix, Eq. (10) does obtain.

$$\mathbf{q} = \mathbf{A}\boldsymbol{\beta}, \mathbf{A} = \begin{bmatrix} 1 & 0 \\ 0 & 1 \end{bmatrix}, \boldsymbol{\beta} = \begin{bmatrix} q_1(t) \\ q_2(t) \end{bmatrix} \tag{10}$$

The Eq. (9) is rewritten to get Eq. (11):

$$\mathbf{T} = \mathbf{T}_{\boldsymbol{\beta}=0} + \mathbf{Z}\boldsymbol{\beta}, \boldsymbol{Z} = \boldsymbol{XA} \tag{11}$$

Therefore, the target equation was modified as follows to obtain Eq. (12):

$$S = (\mathbf{Y} - \mathbf{T})^T (\mathbf{Y} - \mathbf{T}) \tag{12}$$

Substituting Eq. (11) into Eq. (12) results in the Eq. (13):

$$S = \left(\mathbf{Y} - \mathbf{T}_{\boldsymbol{\beta}=0} + \mathbf{Z}\boldsymbol{\beta}\right)^T \left(\mathbf{Y} - \mathbf{T}_{\boldsymbol{\beta}=0} + \mathbf{Z}\boldsymbol{\beta}\right) \tag{13}$$

Therefore, when performing the minimization operation of Eq. (13) concerning $\boldsymbol{\beta}$, the relationship defined by Eq. (14) is deduced to estimate the unknown boundary conditions:

$$\hat{\boldsymbol{\beta}} = \left(\mathbf{X}^T\mathbf{Z}\right)^{-1}\mathbf{Z}^T\left(\mathbf{Y} - \mathbf{T}_{\boldsymbol{\beta}=0}\right) \tag{14}$$

The development of the algebraic operations indicated by Eq. (14) determines Eqs. (15)–(17) as the applicable relations for any set of thermal histories and the parameter value of the number of future time steps for stabilizing the solution.

$$\hat{\boldsymbol{\beta}} = \begin{bmatrix} \hat{q}_1(M) \\ \hat{q}_2(M) \end{bmatrix} = \frac{1}{(C_{11}C_{22} - C_{12}C_{12})} \begin{bmatrix} C_{22}D_1 - C_{12}D_2 \\ -C_{12}D_1 + C_{11}D_2 \end{bmatrix} \tag{15}$$

$$C_{uv} = \sum_{l=1}^{\ell} \sum_{j=1}^{m} \left[\sum_{p=1}^{l} a_{ju}(p)\right] \left[\sum_{p=1}^{l} a_{jv}(p)\right], u = 1, 2; v = 1, 2 \tag{16}$$

$$D_w = \sum_{l=1}^{\ell} \sum_{j=1}^{m} \left[\prod_{p=1}^{l} a_{jw}(p)\right] \left[\mathbf{Y}_{j,M+p-1} - \mathbf{T}_{j,\boldsymbol{\beta}(M+p-1)=0}\right], w = 1, 2 \tag{17}$$

Where the terms a_{ju}, a_{jv}, a_{jw} are components of the matrices, which in turn are the components of the sensitivity coefficient matrix, which is defined by Eq. (18):

$$a_{jK}(p) = \frac{\partial T\left(\mathbf{x}j, t_p\right)}{\partial\left[q_K(1)\right]} \tag{18}$$

The field of sensitivity coefficients does calculate by the same procedure used for the example in the previous section. Therefore, it is necessary to formulate the equivalent problem expressed in **Figure 4**. From Eq. (18), it can be deduced that the term $\mathbf{x}_j$ refers to the histories of the sensitivity coefficients corresponding to the positions of the sensors used to record the thermal histories $\mathbf{Y}_j$.

According to the above, **Figure 12** summarizes the graphs of the analysis of heat extraction from a 50.8 mm external diameter cylinder with a 12.7 mm bore. The hollow cylinder was heated to a temperature of 150°C and exposed to room temperature airflow through the central duct and to still air on the outer surface. The holes to instrument the hollow cylinder near its thermally active borders met the same conditions as in the case presented in Section 2.1 (1 mm diameter at 1.6 mm distance from the surfaces).

Figure 12a and **b** show the recorded cooling curves $Y_j(t)$ (open circle markers) and the calculated $\hat{Y}_j(t)$ (solid lines) with the DHCP solution using the boundary conditions thermals estimated by the IHCP solution. **Figure 12c** and **d** show the functions $q_1(t)$ and $q_2(t)$ in terms of heat flux and heat flow to note that the magnitude of the surface may indicate a misunderstanding on which boundary is more heat extracted. Initially, the most significant amount of heat does remove through the inner surface, where a forced convective phenomenon occurs experimentally.

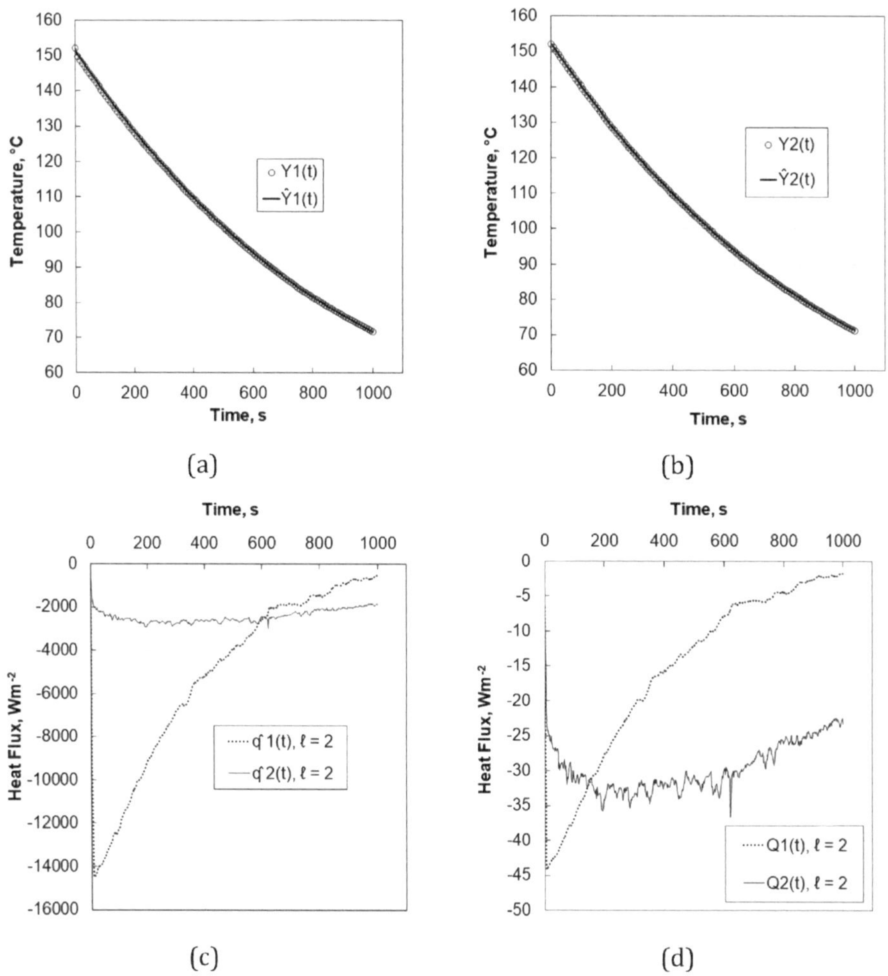

Figure 12.
Comparison of recorded thermal histories $Y_j(t)$ and calculated $\hat{Y}_j(t)$ at (a) inner boundary and (b) outer boundary of the hollow cylinder system, and estimated thermal boundary conditions defined as (c) heat flux $\hat{q}(t)$ and (d) heat flow $Q(t)$, respectively. Figure prepared by the authors.

2.3 Bidimensional analysis of heat flow in radial and axial directions: *Ill-posed* problem for estimating multiple functions in one frontier

Another case study from which it can be deduced that applying the IHCP solution for estimating boundary conditions can extrapolate to more complex systems will present below. Again, the characteristics of the system are cited to define the governing heat transfer equation:

- Consider a disk of small thickness where the characteristics of symmetric conditions that determine the gradients in the angular direction of the system are not significant. Nevertheless, a heterogeneously distributed boundary condition does determine from its geometric center on one of the faces with the largest surface. In contrast, on the opposite face, it is considered homogeneous and known.

- The heat conduction in the solid does determine by ignoring the discontinuity in the material due to the insertion of temperature sensors (instrumentation with thermocouples), and the absence of sources or internal heat sinks in the solid does consider.

- The material is considered isotropic, and the thermophysical properties (k, ρ, and Cp) are a function of temperature.

- The heterogeneously distributed boundary condition can be specified by a series of discrete heat flux histories bounded by specific adjacent annular regions of defined width.

- A homogeneous heat flux history can specify the boundary condition on the opposite surface.

- The initial temperature field is known.

- At least one point each region information is available on the temperature distribution inside the body to adequately estimate the unknown thermal boundary conditions.

Figure 13 shows the schematization of the system of interest and the corresponding mathematical formulation.

In this analysis, the mathematical treatment of one of the two previous cases can be implemented to obtain the solution of the IHCP through the objective function. In this case, the methodology followed in Section 2.1 will be chosen. It is worth mentioning that in the case of Section 2.2, the distribution matrix of the domain of the functions on the boundaries of the system must formulate. **Figure 14** and Eq. (19) determine the discretization of the disc in two dimensions and the type of penta-diagonal matrix of the system of equations, respectively. However, the system can increase according to the product (number of nodes in the r-direction) * (number of nodes

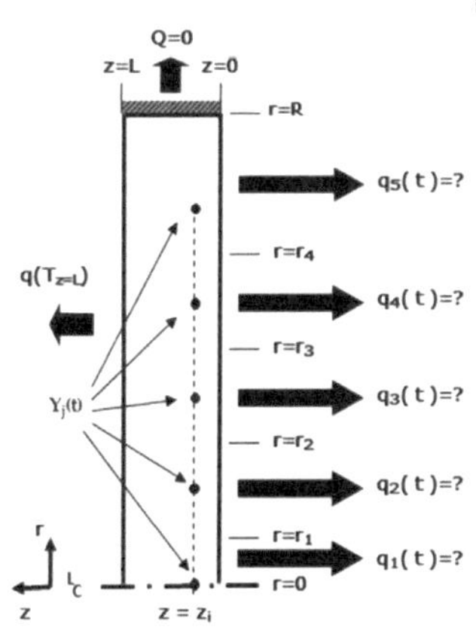

Mathematical model

$$\frac{1}{r}\frac{\partial}{\partial r}\left(rk\frac{\partial T}{\partial r}\right)+\frac{\partial}{\partial z}\left(\frac{\partial T}{\partial z}\right)=\rho C_p\frac{\partial T}{\partial t}$$

Boundary conditions

$$-k\frac{\partial T}{\partial r}=q_\beta(t)=\begin{cases}q_1(t), & 0\le r\le r_1\\ q_2(t), & r_1\le r\le r_2\\ q_3(t), & r_2\le r\le r_3,\\ q_4(t), & r_3\le r\le r_4\\ q_5(t), & r_4\le r\le r_5\end{cases}\qquad z=0,\qquad t>0$$

$$-k\frac{\partial T}{\partial r}=q,\qquad 0\le r\le R,\qquad z=L,\qquad t>0$$

$$\frac{\partial T}{\partial r}=0,\qquad 0\le z\le L,\qquad r=0,\qquad t>0$$

$$\frac{\partial T}{\partial r}=0,\qquad 0\le z\le L,\qquad r=R,\qquad t>0$$

Initial condition

$$T(r,z,t)=T_0,\qquad 0\le r\le R,\qquad 0\le z\le L,\qquad t=0$$

Measured thermal histories

$$Y_j(t),\qquad r=r_j,\qquad z=z_i,\qquad t>0,\qquad j=1..5$$

Figure 13.
Formulation of the IHCP for a cylindrical system with the bidirectional flow in the radial and axial directions with two active boundaries.

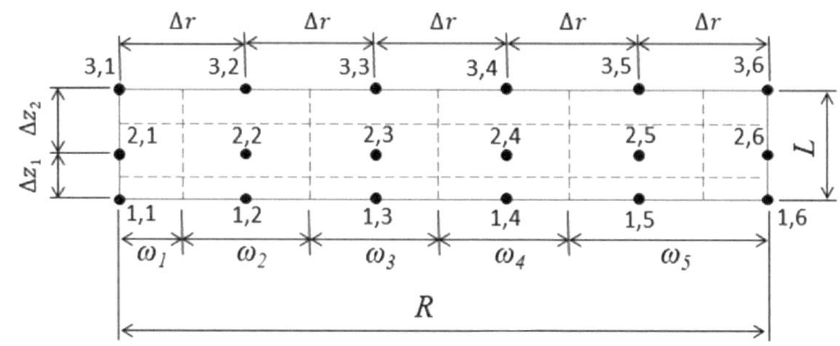

Figure 14.
Simplified scheme of the discretization in two dimensions (r, z) to define the CVs in the disk system.

in the z-direction). Again, matrix solution techniques are recommended by LU decomposition or successive over-relaxation (SOR).

$$\begin{bmatrix} b_{11} & c_{12} & 0 & 0 & 0 & 0 & e_{21} & 0 & 0 & 0 & 0 & 0 & 0 & \cdots \\ a_{11} & b_{12} & c_{13} & 0 & 0 & 0 & 0 & e_{22} & 0 & 0 & 0 & 0 & 0 & \cdots \\ 0 & a_{12} & b_{13} & c_{14} & 0 & 0 & 0 & 0 & e_{23} & 0 & 0 & 0 & 0 & \cdots \\ 0 & 0 & a_{13} & b_{14} & c_{15} & 0 & 0 & 0 & 0 & e_{24} & 0 & 0 & 0 & \cdots \\ 0 & 0 & 0 & a_{14} & b_{15} & c_{16} & 0 & 0 & 0 & 0 & e_{25} & 0 & 0 & \cdots \\ 0 & 0 & 0 & 0 & a_{15} & b_{16} & 0 & 0 & 0 & 0 & 0 & e_{26} & 0 & \cdots \\ d_{11} & 0 & 0 & 0 & 0 & 0 & b_{21} & c_{22} & 0 & 0 & 0 & 0 & e_{31} & \cdots \\ 0 & \ddots & 0 & 0 & 0 & 0 & \ddots & \ddots & \ddots & 0 & 0 & 0 & 0 & \ddots \end{bmatrix}$$

$$\begin{bmatrix} T_{11}^{t} \\ T_{12}^{t} \\ T_{13}^{t} \\ T_{14}^{t} \\ T_{15}^{t} \\ T_{16}^{t} \\ T_{21}^{t} \\ T_{22}^{t} \\ T_{23}^{t} \\ T_{24}^{t} \\ T_{25}^{t} \\ T_{26}^{t} \\ T_{31}^{t} \\ \vdots \end{bmatrix} = \begin{bmatrix} T_{11}^{t+\Delta t} - f_{11} \\ T_{12}^{t+\Delta t} - f_{12} \\ T_{13}^{t+\Delta t} - f_{13} \\ T_{14}^{t+\Delta t} - f_{14} \\ T_{15}^{t+\Delta t} - f_{15} \\ T_{16}^{t+\Delta t} - f_{16} \\ T_{21}^{t+\Delta t} - f_{21} \\ T_{22}^{t+\Delta t} - f_{22} \\ T_{23}^{t+\Delta t} - f_{23} \\ T_{24}^{t+\Delta t} - f_{24} \\ T_{25}^{t+\Delta t} - f_{25} \\ T_{26}^{t+\Delta t} - f_{26} \\ T_{31}^{t+\Delta t} - f_{31} \\ \vdots \end{bmatrix} \quad (19)$$

As previously defined, the objective function does minimize concerning the set of estimable heat flux density functions $\mathbf{q}$, such that Eq. (20) is determined:

$$\frac{\partial S}{\partial \mathbf{q}} = \frac{\partial}{\partial \mathbf{q}}\left[\sum_{l=1}^{\ell}\sum_{j=1}^{m}\left(Y_{j,M+l-1} - T_{j,M+l-1}\right)^2\right] = 0 \tag{20}$$

The procedure leads to the expression of the Eq. (21):

$$0 = \sum_{l,j}\left(Y_{j,M+l-1} - T_{j,M+l-1}\right)\frac{\partial T_{j,M+l-1}}{\partial q_{\beta,M}}, \beta = 1, 2, \ldots, 5 \tag{21}$$

A general discrete definition of the Taylor series expansion for the determination of the thermal field does express as (Eq. (22)):

$$T_M = T_{M-1} + f\left[T_{M-1}, q_{M-1}\right]\Delta q \tag{22}$$

The differential form of Eq. (22) applied to the case study can be expressed as Eq. (23):

$$T_{j,M+l-1} = T^*_{j,M+l-1} + \sum_{l,j}\left(\frac{\partial T^*_{j,M+l-1}}{\partial q_{\beta,M}}\right)\left(q_{\beta,M} - q_{\beta,M-1}\right), \beta = 1, 2, \ldots, 5 \tag{23}$$

By substituting Eq. (22) in Eq. (20), the expression for the solution of the IHCP obtain Eq. (24):

$$\sum_{\alpha=1}^{n}\left(q_{\beta,M} - q_{\beta,M-1}\right)\sum_{l,j}\left(\frac{\partial T^*_{j,M+l-1}}{\partial q_{\beta,M}} \bullet \frac{\partial T^*_{j,M+l-1}}{\partial q_{\beta,M}}\right) =$$

$$\sum_{l,j}\left(Y_{j,M+l-1} - T^*_{j,M+l-1}\right)\frac{\partial T^*_{j,M+l-1}}{\partial q_{\beta,M}}, \beta = 1, 2, \ldots, 5 \tag{24}$$

Eqs. (22) and (23) again show the definition of the sensitivity coefficient. The data set corresponding to the histories at the position of the jth sensor due to a change in the heat flux in the βth domain of the boundary at the computation time M defines the equivalent of the cooling rate histories recorded at those points of the solid. As can be deduced, each set is independent and isolated in the boundary condition over the domain defined in front of the sensor at the time of calculation without considering changes in the other boundary conditions.

Figure 15 shows the results of a benchmark case to complete the exposition. **Tables 2–6** summarize the information on the direct and inverse problem statements in **Figure 13**. The material is considered AISI 304 stainless steel; therefore, the average values of the thermophysical properties of this material did use. The analyzed disk has dimensions of 100 mm in diameter and a thickness of 6 mm.

The five functions that will distribute in the annular regions of the disk surface defined in **Table 2** do consider to develop the case of analysis. The functions are determined as histories of heat flux density and initially present a constant period, later to reach a maximum value with a steep slope and then return to the same state of constant magnitude. Each function presents an initial period of different tendency such that the curves $q_i(t)$ overlap at some instant, as shown in **Figure 15a** and the relationships in **Table 6**. Formulating the DHCP concerning the system in **Figure 13**

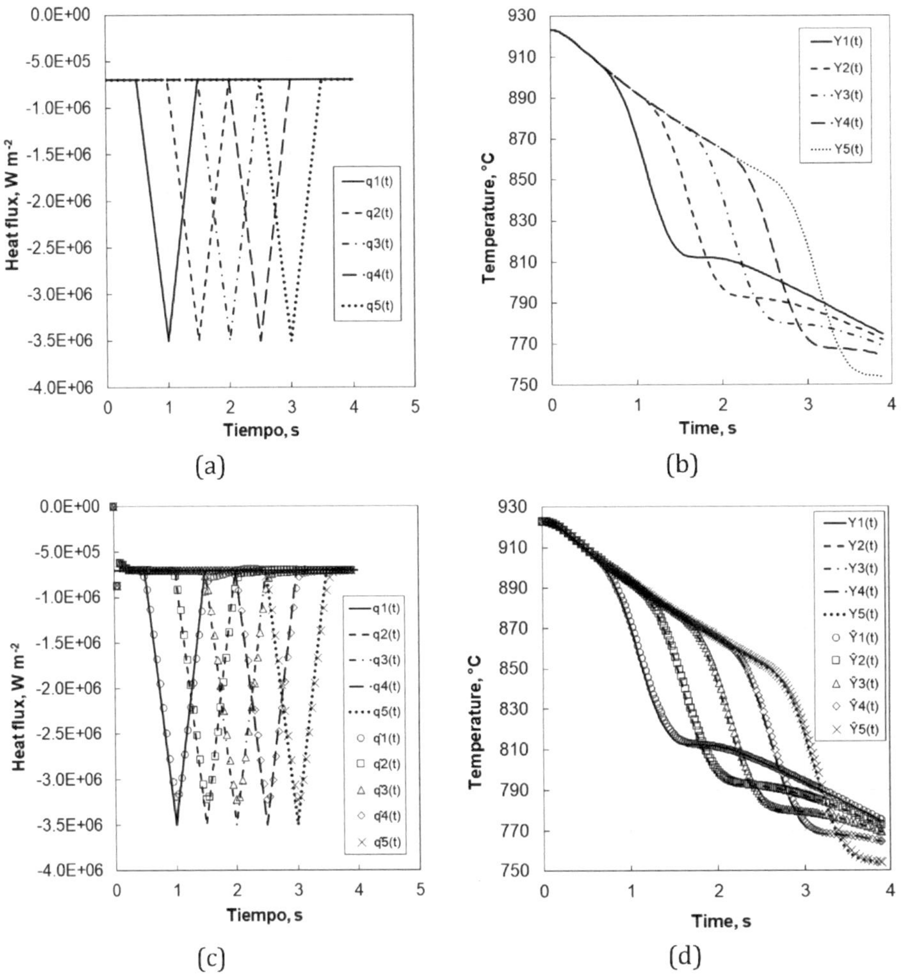

Figure 15.
Benchmark case for the analysis of the multiple estimations of heterogeneously distributed thermal boundary conditions over the larger-magnitude surfaces of a disk; (a) the proposed $q_i(t)$ functions do virtually distribute in ω_i sections, and DHCP is solved to obtain (b) the thermal histories $Y_i(t)$ at virtual sensors positions. In turn, these do use to solve the IHCP estimating the thermal boundary conditions $\hat{q}_i(t)$, and (c) they are compared with the origin functions, and finally (d) the thermal histories $\hat{Y}_i(t)$ originated by a new DHCP solution are compared in this cycle.

ω_1	ω_2	ω_3	ω_4	ω_5
$0 \leq r < 0.005$	$0.005 \leq r < 0.015$	$0.015 \leq r < 0.025$	$0.025 \leq r < 0.035$	$0.035 \leq r \leq 0.05$

Table 2.
Dimensions of the bounded regions for the specification of the heterogeneous boundary condition ($0 \leq r < R$).

Y_1	Y_2	Y_3	Y_4	Y_5
(0, 0.0018)	(0.01, 0.0018)	(0.02, 0.0018)	(0.03, 0.0018)	(0.04, 0.0018)

Table 3.
Ordered pair (r, z) to define the location of the temperature sensors $\mathbf{Y}_j(t)$.

Material properties	k (Wm^{-1}°C^{-1})	$\rho \bullet Cp$ (Jm^{-3}°C^{-1})
	22.0	5.125×10^6

Table 4.
Average thermo-physical properties of AISI 304 type stainless steel.

CV_r	CV_z	SOR parameter	ℓ
50	10	0.6	3

Table 5.
Parameters for the numerical solution.

ω_1	$q_1(t)$	$-7.0E5\ 0.0 \le t \le 0.5$ $-5.6E6\,t + 2.1E6\ 0.5 < t < 1.0$ $5.6E6\,t - 9.1E6\ 1.0 \le t \le 1.5$ $-7.0E5\ 1.5 \le t \le 4.0$
ω_2	$q_2(t)$	$-7.0E5\ 0.0 \le t \le 1.0$ $-5.6E6\,t + 4.9E6\ 1.0 < t < 1.5$ $5.6E6\,t - 1.19E6\ 1.5 \le t \le 2.0$ $-7.0E5\ 2.0 \le t \le 4.0$
ω_3	$q_3(t)$	$-7.0E5\ 0.0 \le t \le 1.5$ $-5.6E6\,t + 7.7E6\ 1.5 < t < 2.0$ $5.6E6\,t - 1.47E6\ 2.0 \le t \le 2.5$ $-7.0E5\ 2.5 \le t \le 4.0$
ω_4	$q_4(t)$	$-7.0E5\ 0.0 \le t \le 2.0$ $-5.6E6\,t + 1.05E6\ 2.0 < t < 2.5$ $5.6E6\,t - 1.75E6\ 2.5 \le t \le 3.0$ $-7.0E5\ 3.0 \le t \le 4.0$
ω_5	$q_5(t)$	$-7.0E5\ 0.0 \le t \le 2.5$ $-5.6E6\,t + 1.33E6\ 2.5 < t < 3.0$ $5.6E6\,t - 2.03E6\ 3.0 \le t \le 3.5$ $-7.0E5\ 3.5 \le t \le 4.0$

Table 6.
Parameters of the distributed heterogeneous boundary condition.

defines the boundary condition of the opposite with a value $q(t) = 0$ and an initial field of constant temperatures of 920°C.

The solution with the discretization proposed in **Table 5** and with the position of the sensors defined in the places indicated in **Table 3** is shown graphically in **Figure 15b** with cooling curves $Y_i(t)$. It can be noted that the system cools a maximum of 170°C and that the gradients generated in the mass of each region define the subsequent cooling trend of the entire system. It can also be noted that neither the sharp change points of the $q_i(t)$ original functions nor the maximums of each curve are accurately estimated with the corresponding $\hat{q}_i(t)$ (see **Figure 15c**). However, when solving the DHCP with these new boundary conditions, a very acceptable adjustment of the $\hat{Y}_i(t)$ cooling curves is observed (see **Figure 15d**), so the differences between these systems with the approximation may be due to the approximation criteria implied by the numerical methods for both the DHCP and IHCP solution.

2.4 Applications

One of the most impressive fields of application of inverse analysis in heat conduction problems does link to the aerospace industry, specifically in the race for mastery of trips to outer space. The knowledge of the heating conditions of the surfaces of the shuttles exposed to the situation of reentry into the atmosphere contributed to the design of coatings, selection of suitable materials, and cooling systems to counteract the extreme heating and consequent failures. In this way, the same concept is applied to systems at the level of the earth's surface when considering nuclear fission and fusion reactors. In other processes of interest, the analysis is projected to the design and monitoring of reactors for extracting mineral substances and their subsequent chemical refining, as well as in the design processes of materials with optimal allotropic properties, either by sintering or thermosetting, where the control of the rate of temperature change is depended on to obtain optimal microstructural characteristics. In general, one can think of any other system where thermal boundary conditions are complex to determine by fluid dynamics analysis, radiant systems, or a combination thereof. However, a limitation occurs with the impossibility of instrumenting with sensors and implementing data acquisition systems in the advice of interest. Another stemmed from the ability of computer systems to process a large amount of data, which also determined the number of dimensions implicit in the analyses. However, at present, it has been minimized. The selection of the number of analysis dimensions will depend on the degree of detail that defines the characteristics of the boundary condition and the conditions of the phenomenon studied.

3. Conclusions

It is shown in the case studies that it is possible to analyze the phenomena of heat transfer by conduction in solids by formulating inverse problems using essential foundations of mathematical concepts. As it has been talked about in the course of the discussion, treating these problems requires an appropriate quality of information that will process to determine thermal boundary conditions as a primary objective. Other complications associated with the stability of the solution to the *ill-posed* problem can adopt different theories. However, the sequential process of estimating the function by implementing the least square's objective function guarantees an efficient technique, such that only at least one recorded thermal history near the boundary of interest is needed to obtain an estimate. More invasive instrumentation would significantly affect the continuity of the materials. Therefore, with the knowledge bases about the phenomena that occur, correct criteria can be taken regarding the precision of the IHCP solution. In such a way that, for example, it can deduce that the analyses are highly reliable due to finding magnitudes of relatively high gradients in the systems under study. Estimating thermal boundary conditions through the IHCP solution helps to weigh the magnitudes associated with interaction with highly complex external conditions. Additionally, it is practical to define the boundary conditions of transient systems in terms of heat flux histories due to the state of the mathematical formulation that governs the problems. The information can be used as a validation parameter for modeling convective, radiant, or a combination of both media transfer phenomena. The amount of data currently represents a manageable setback due to the high capacity of computer systems.

Acknowledgements

Thanks are due to the Tecnológico Nacional de México, which, through the Instituto Tecnológico de Morelia, provided financial support for the publication of this material.

Conflict of interest

The authors declare no conflict of interest.

Nomenclature

β	index of the heat flux function
$\boldsymbol{\beta}$	heat flux function matrix
$\hat{\boldsymbol{\beta}}$	matrix of estimated functions of heat flux
∂	partial derivative
Δ	magnitude of increase
	thermal histories data index
ρ	density
θ	angular coordinate
ω_i	radial domain magnitude
a_i	coefficient of systems of algebraic equations
A	area perpendicular to the direction of unidirectional flow
$\mathbf{A}$	boundary conditions individuality matrix
$a_{j\text{u}}$	component of the sensitivity coefficient matrix matrices
$a_{j\text{v}}$	component of the sensitivity coefficient matrix matrices
$a_{j\text{w}}$	component of the sensitivity coefficient matrix matrices
b_i	coefficient of systems of algebraic equations
c_i	coefficient of systems of algebraic equations
C_p	specific heat capacity
C_uv	IHCP solution component
d_i	coefficient of systems of algebraic equations
D_w	IHCP solution component
e_i	coefficient of systems of algebraic equations
f_i	coefficient of systems of algebraic equations
$\mathbf{F}$	function of the Dirichlet, Neumann or Robinson type boundary condition
g_i	coefficient of systems of algebraic equations
h	convective heat transfer coefficient
i	geometric discretization index/index of mathematical operators
j	sensor index/index of mathematical operators
k	thermal conductivity
l	index of the number of future time steps
ℓ	number of future time steps
m	number of sensors
M	reference to the calculation time instant
n	number of control volumes in radial direction
N	number of data from thermal histories

p	index of mathematical operators
q	heat flux function
$\hat{q}$	estimated heat flux function
$\mathbf{q}$	heat flux function matrix
$\dot{q}$	heat generated or consumed per unit volume
Q	heat flow
r	radial coordinate
R	magnitude of the radius of cylindrical systems
S	least squares function (objective function)
S_Y	standard deviation from thermal histories
t	variable time
T	variable temperature
T_s	surface temperature
T_∞	bulk fluid temperature
T^*	abbreviation for sensitivity coefficient definition
$\mathbf{T}$	temperature field
u	index of mathematical operators
v	index of mathematical operators
V_i	amount of control volume
w	geometric discretization index/index of mathematical operators
$\mathbf{x}_j$	histories of sensitivity coefficients
$\mathbf{X}$	sensitivity coefficient matrix
Y	thermal history
$\mathbf{Y}$	thermal history matrix
z	axial coordinate

Author details

Jorge Sergio Téllez-Martínez*, Miriam Zulma Sánchez-Hernández,
María Janeth Vega-Flores, Abel Alberto Pintor-Estrada, Hugo Enrique Alva-Medrano
and Nicolás David Herrera-Sandoval
Tecnológico Nacional de México/Instituto Tecnológico de Morelia, Morelia, México

*Address all correspondence to: jorge.tm@morelia.tecnm.mx

References

[1] Dowding KJ, Beck JV. A sequential gradient method for the inverse heat conduction problem (IHCP). Journal of Heat Transfer – Transactions of ASME. 1999;**121**:300-306

[2] Beck JV, Blackwell B, Clair CS. Inverse Heat Conduction: Ill Posed Problems. New York: Wiley; 1985

[3] Beck JV, Blackwell B, Haji-Sheikh A. Comparison of some inverse heat conduction methods using experimental data. International Journal of Heat and Mass Transfer. 1996;**39**(17):3649-3657

[4] Hernández-Morales B, Cruces-Reséndez R, Téllez-Martínez JS. Revisiting the temperature gradient method. Materials Performance and Characterization. 2018;**8**(2):170-187. DOI: 10.1520/MPC20180031

[5] Meekisho L, Hernandez-Morales B, Tellez-Martinez JS, Chen X. Computer-aided cooling curve analysis using WinProbe. International Journal of Materials and Product Technology. 2005; **24**(1-4):155-169. DOI: 10.1504/IJMPT.2005.007946

[6] Krzysztof G. Inverse heat conduction problems. In: Vikhrenko V, editor. Heat Conduction - Basic Research. London, United Kingdom: IntechOpen [Online]; 2011. DOI: 10.5772/26575

[7] Xue X, Luck R, Berry JT. Comparisons and improvements concerning the accuracy and robustness of inverse heat conduction algorithms. Inverse Problems in Science and Engineering. 2005;**13**:177-199

[8] Dowding KJ. Multi-dimensional analysis of quenching: comparison of inverse techniques. In: Proceedings of the 18th Heat Treating Conference, 12-15 October 1998, Rosemont; Illinois. pp. 525-534

[9] Blanc G, Raynaud M. The Hiep Chau. A guide for the use of the function specification method for 2D inverse heat conduction problems. Revue Générale de Thermique. 1998;**37**(1):17-30

[10] Prasanna Kumar TS. A serial solution for the 2-D inverse heat conduction problem for estimating multiple heat flux components. Numerical Heat Transfer Part B-Fundamentals. 2004; **45**(6):541-563

[11] Fan C, Sun F, Yang L. A simple method for inverse estimation of surface temperature distribution on a flat plate. Inverse Problems in Science and Engineering. 2009;**17**(7):885-899

[12] Woodfield PL, Monde M, Mitsutake M. Implementation of analytical two-dimensional inverse heat conduction technique to practical problems. International Journal of Heat and Mass Transfer. 2006;**49**:187-197

[13] Murio DA, Mejia CE, Zhan S. Discrete mollification and automatic numerical differentiation. International Journal of Computers and Mathematics with Applications. 1998; **35**(5):1-16

[14] Shenghe Zhan C, Coles DAM. Automatic numerical solution of generalized 2-D IHCP by discrete mollification. International Journal of Computers and Mathematics with Applications. 2001;**41**(1):15-38

[15] Qian Z. An optimal modified method for a two-dimensional inverse heat conduction problem. Journal of Mathematical Physics. 2009;**50**(2): 023502-1-023502-9

[16] Nallathambi AK, Spetch E. Estimation of the heat flux in array of jets quenching using experimental and inverse finite element method. Journal of Material Processing Technology. 2009; **209**:5325-5332

[17] Prasanna Kumar TS, Kamath HC. Estimation of multiple heat-flux components at the metal/mold interface in bar and plate aluminum alloy castings. Metallurgical and Materials Transactions B. 2004;**35**(3):575-585

[18] Kamath HC, Prasanna Kumar TS. Multidimensional analysis of interface heat flux in metallic molds during solidification aluminum alloy plate castings. In: 108th Casting Congress of American Foundrymen's Society, June 2004. Rosemont; Illinois, USA

[19] Prasanna Kumar TS, Pareek A, Arjun N. Determination of heat transfer coefficient during quenching of an alloy steel by in-situ plant testing. In: Proceedings of 6th International Conference on Quenching and Distortion Control, 9-14 September 2012, Chicago; Illinois

Chapter 2

Heat Transfer Fundamentals Concerning Quenching Materials in Cold Fluids

Nikolai Kobasko

Abstract

The current chapter discusses three principles of heat transfer related to transient nucleate boiling process when any film boiling during quenching is completely absent. They include the duration of nucleate boiling establishment, surface temperature behavior during self-regulated thermal process, and generalized fundamental equation for evaluation of transient nucleate boiling length. In fact, majority of authors in their analytical and experimental investigations always considered three classical heat transfer modes: film boiling, nucleate boiling, and convection. It is shown in the chapter that the absence of film boiling process simplifies numerical and analytical calculations. It is very important to know that mentioned principles can be used for calculation of temperature fields and stresses and development of new technologies without performing costly experiments. It is stated that universal correlation for heating and cooling time evaluation, modified by proposed principles, can be widely used for recipes development when exploring new intensive quenching technologies. Examples of calculations are provided.

Keywords: film boiling absence, three principles, heat transfer, new approach, recipes development, new technologies

1. Introduction

This chapter considers recently discovered unusual characteristics of transient nucleate boiling process. It was possible to do that terminating any film boiling process during quenching. Since it was a widely distributed opinion on three stages of cooling (film boiling, nucleate boiling, and convection), scientists paid the main attention to studying film boiling processes. The results of investigation obtained by scientists are really very important and very interesting since they allowed to decrease distortion. Four types of heat transfer modes discovered [1, 2] were important for the practice. It was established that vapor film behavior depends on size and form of the quenched steel part. These accurate investigations brought success and lifted the heat-treating technology to the next level in progressing scale. Nobody considered cooling curves during quenching without considering the film boiling process because existing theory and accurately performed experiments in many cases showed vapor films during quenching in liquid media. Engineers used this concept in practice. Even thermal scientists were sure that film boiling during hardening steel in cold fluids always exists. However, accurate experiments of French [3], which were performed in cold 5% water slow agitated solution of sodium hydroxide (NaOH), clearly displayed the absence of film boiling process when quenching spherical steel samples of different diameters from 875°C in the mentioned fluid. Accurate experiments of French were published in many papers and books [3]. Scientists did not discuss them and did not use them for solving the inverse problem to investigate heat transfer when any film boiling process is completely absent during quenching. Probably, it was impossible to explain the absence of film boiling process during quenching probes from 875°C in a slow agitated cold fluid. The current short overview presents results of investigations which were obtained using long-lasting experiments of different authors including experiments of French. As a result, the three of the most important principles were formulated, which create the fundament for intensive quenching processes. These new principles are discussed in the current short chapter.

2. Incorrect heat transfer coefficient evaluation during transient nucleate boiling

Historically in the heat-treating industry, the heat transfer coefficient (HTC) during transient nuclear boiling process is evaluated as $\alpha_{eff} = q/(T - T_m)$, which is called the effective HTC. In fact, the real HTC is evaluated as $\alpha_{real} = q/(T - T_s)$. As known, the critical radius R_{cr} of a bubble growth depends on the overheat of a boundary layer $\Delta T_S = T_{sf} - T_S$, which is determined as [4]:

$$R_{cr} = \frac{2\sigma T_S}{r^* \rho'' \Delta T_S}, \tag{1}$$

Note that R_{cr} is critical radius of a bubble that is capable of growing and functioning; σ is surface tension (N/m); T_S is saturation (boiling) temperature; r* is latent heat of evaporation (J/kg); ρ'' is vapor density (kg/m^3); and $\Delta T_S = T_{sf} - T_S$ is wall overheat. HTC related to difference $\Delta T_S = T_{sf} - T_S$ is known as a real HTC and it is called unreal (effective) if it is related to $\Delta T_m = T_{sf} - T_m$. Here, T_{sf} is wall temperature and T_m is bath temperature. Effective HTC is used only for approximate core cooling rate and core cooling time evaluation and cannot be used for correct temperature field calculation. There is a big difference between the real and effective HTCs. **Table 1** shows the difference between real and effective HTCs depending on bath temperatures T_m.

As seen from **Table 1**, effective HTC is decreased up to 700%, as compared with the real HTC. That is why the cooling process during transient nucleate boiling was considered as the slow cooling that requires the use of powerful pumps or powerful rotating propellers to make the cooling process more intensive via strong agitation of the quenchant.

Temperature T_m, ^oC	15	20	30	40	60	80	90	95
$\alpha_{real}/\alpha_{eff}$	7.07	6.71	6.00	5.29	3.86	2.43	1.71	1.26
$\alpha_{real}/\alpha_{eff}$, %	707	671	600	529	386	243	171	126

Table 1.
Ratio of real HTC to effective HTC versus temperature of water.

3. Finite initial heat flux density to be compared with its critical value q_{cr1}

It was accepted by worldwide community that during quenching steel from high temperatures in cold fluid, three stages of cooling are always present: film boiling, transient nucleate boiling, and convection. It is supported by the conventional law of Fourier (2):

$$q = -\lambda \frac{\partial T}{\partial r} \tag{2}$$

During immersion of a heated steel part into cold liquid, at the very beginning of cooling $q \to \infty$. It means that at the very beginning, initial heat flux density always accedes to the first critical heat flux density q_{cr1}, resulting in developed film boiling process. However, Lykov [5] considered modified law of Fourier (3) resulting in finite heat flux density:

$$q = -\lambda \frac{\partial T}{\partial r} - \tau_r \frac{\partial T}{\partial \tau} \tag{3}$$

It was shown by him that modified law of Fourier generates hyperbolic heat conductivity eq. (4) [5]:

$$\frac{\partial T}{\partial \tau} + \tau_r \frac{\partial^2 T}{\partial \tau^2} = a div(gradT) \tag{4}$$

Here, λ is thermal conductivity of solid material in W/mK and τ_r is relaxation time.

For the first time, hyperbolic heat conductivity Eq. (4) with the appropriate boundary and initial conditions was solved analytically by authors [6] and it was shown that initial heat flux density is finite value which can be far below the first critical heat flux density q_{cr1}. It means that the film boiling process is absent completely when quenching steel parts from high temperatures in cold fluids. Such conclusion was formulated from the point of view of mathematics. One can formulate the same conclusion from the point of view of physics. Really, at the very beginning of cooling the cold fluid must be heated to the boiling point of a liquid and then overheated to initiate the nucleate boiling process. During this period of time, the surface of the steel part decreases, thereby creating a temperature gradient that results in finite initial heat flux density. At that moment of time, heat flux density is compared with the first critical heat flux density. If it is below q_{cr1}, any film boiling process is completely absent. To support such conclusion, it makes sense to consider experiments of French [3] presented by **Table 2**.

As seen from **Table 2**, surface cooling curves for spherical probes of sizes such as 6.35 mm, 12.7 mm, and 120 mm are practically similar and cooling short time for all of them is almost the same.

Size, Inches, (mm)	Time (s)							
	700°C	600°C	500°C	400°C	300°C	250°C	200°C	150°C
0.25″ (6.35)	0.027	0.037	0.043	0.051	0.09	0.15	0.29	0.69
0.5″ (12.7)	0.028	0.042	0.058	0.071	0.11	0.15	0.26	0.60
4.75″ (120.6)	0.043	0.066	0.09	0.12	0.17	0.21	0.29	0.95

Table 2.
Time required for the surface of steel spheres of different sizes to cool to different temperatures when quenched from 875°C in 5% water solution of NaOH at 20°C agitated with 0.914 m/s [3].

Similar results on drastic decrease of surface temperature, which drops within short time almost to the boiling point of a quenchant, were obtained by different authors (see **Figures 1–3**).

In **Figures 1–3**, the surface temperature of probes after initial drastic decrease maintains at the level of boiling point of a liquid and is called the self-regulated thermal process (SRTP).

Figure 4 presents a comparison of experimental core cooling curve with core cooling curve obtained for constant surface temperature during self -regulated thermal process.

Table 3 presents initial temperature T_I and temperature T_{II} at the end of self-regulated thermal process (SRTP) versus thickness of the stainless probes.

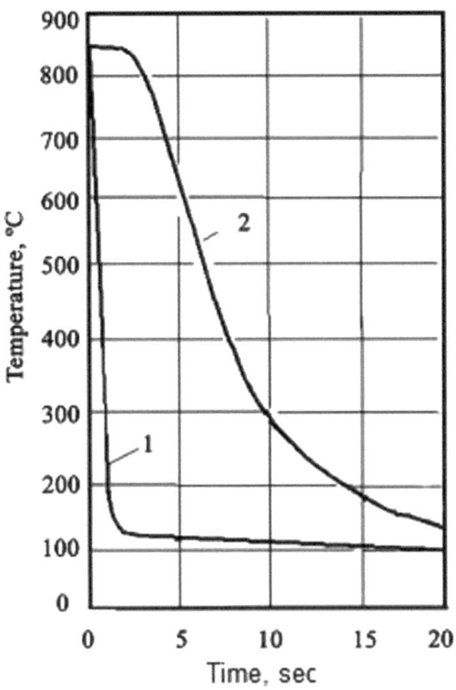

Figure 1.
Surface (1) and core (2) temperature curves versus time during quenching cylindrical probe 20 mm in diameter in low concentration of water polymer solution at 20°C [7].

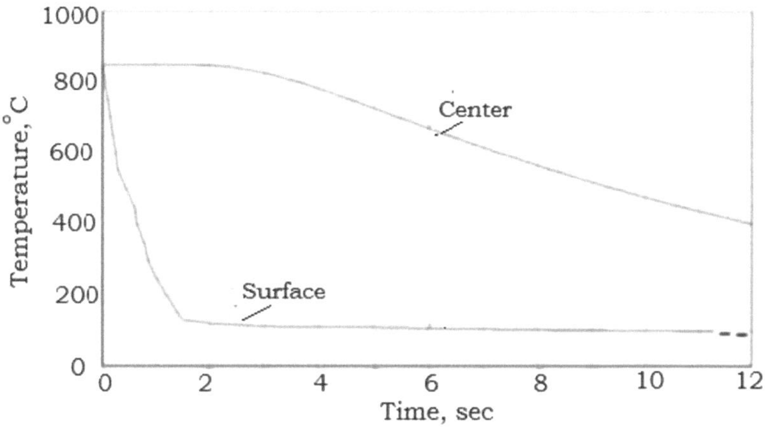

Figure 2.
Cooling curves obtained in Idemitsu Kosan Co., Ltd. Lab (Japan) for cylindrical specimen of 28 mm diameter and 112 mm length when quenching in water flow of 1.5 m/s at 20°C [8].

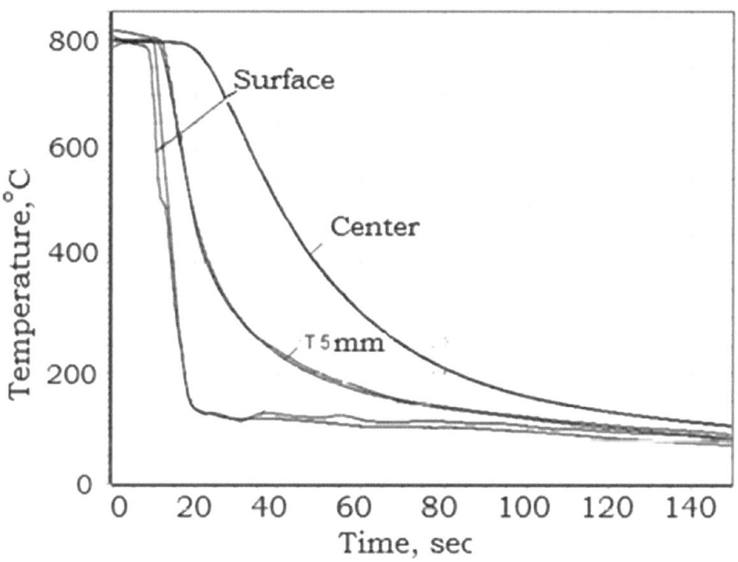

Figure 3.
Cooling curves versus time during quenching cylindrical probe 50 mm diameter in 14% NaCl water solution at 23°C [9].

The average surface temperature for a probe 50 mm in diameter during numerical calculation within the self-regulated thermal process was approximately equal to 114°C (see **Figure 3** and **Table 3**). Error due to averaging of surface temperature is rather small and is equal to −0.47% (see **Table 4**).

Based on unusual characteristics of SRTP, intensive quenching processes IQ-2 and IQ-3 were developed. The IQ-2 technological process explores transient nucleate boiling (see **Figure 5a**), while transient nucleate boiling process is absent while performing IQ-3 technology (see **Figure 5b**). It is very important to know how much technological process IQ-3 differs from technological process IQ-2, which is essentially cheaper.

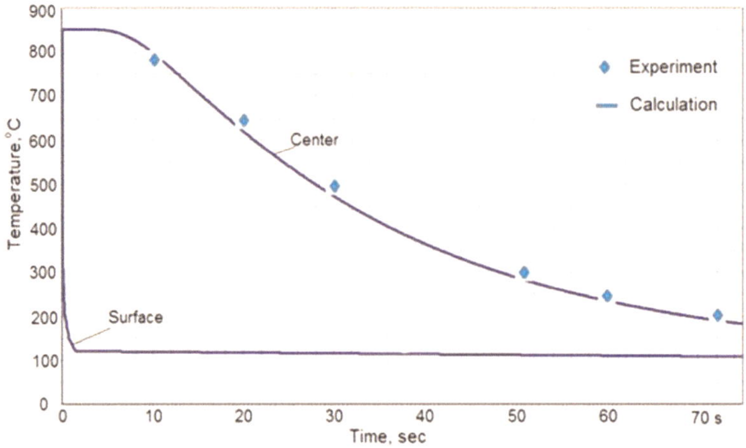

Figure 4.
A comparison of experimental cooling curve with calculated cooling curve when surface temperature during quenching 50 mm probe in water salt solution is constant.

Diameter or thickness in mm	T_I, °C	T_{II}, °C	$(T_I + T_{II})/2$
20	126	107.2	117
40	121	107.2	114
50	120	107.2	113.6
60	119	107.2	113
80	117.3	107.2	112
100	116	107.2	111.5
120	115.4	107.2	111

Table 3.
Initial temperature T_I *and temperature* T_{II} *at the end of self-regulated thermal process (SRTP) versus thickness of the stainless probes.*

τ_{nb}, s	T_{center}, °C experiment	$\overline{T}_{center}$, °C calculation	ε, %
0	850	850	0
5	847.5	847.5	0
10	797.2	797.06	−0.016
15	708.2	707.6	−0.085
20	620	619	−0.16
25	541	539.5	−0.3
30	472.5	470.8	−0.36
35	414	412.3	−0.41
40	364.7	362.9	−0.49
45	323	321.2	−0.56

τ_{nb}, s	T_{center}, °C experiment	$\overline{T}_{center}$, °C calculation	ε, %
50	287.7	286.3	−0.47
55	258	257.23	−0.30
60	233.7	233	−0.30
65	210.8	212.9	+0.99
70	195.7	196.1	+0. 22
75	181	182.5	+0.83

Table 4.
Errors at the core of a cylindrical probe 50 mm diameter generated by replacing the real surface temperature during SRTP with average constant surface temperature.

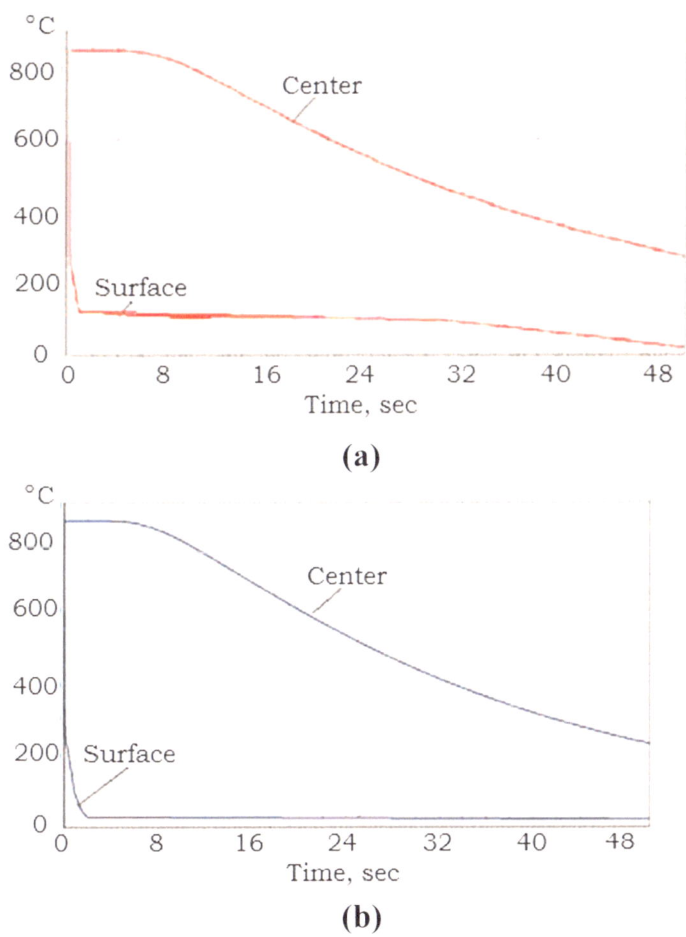

Figure 5.
The temperature difference between IQ-2 and IQ-3 processes when quenching a cylindrical steel probe 50 mm in diameter in an agitated water solution at 20°C and in condition when HTC is infinity: a) it is an IQ-2 process and b) it is a IQ-3 process.

Temperature interval	850–500°C	850–400°C	850–300°C	700–400°C
Cooling time for IQ-3 process in seconds	25.8	32.8	37	18
Cooling time for IQ-2 process in seconds	27.6	36	46.8	20.7

Table 5.
Core cooling time difference between IQ-2 and IQ-3 processes when transient nucleate boiling and direct convection occur during quenching.

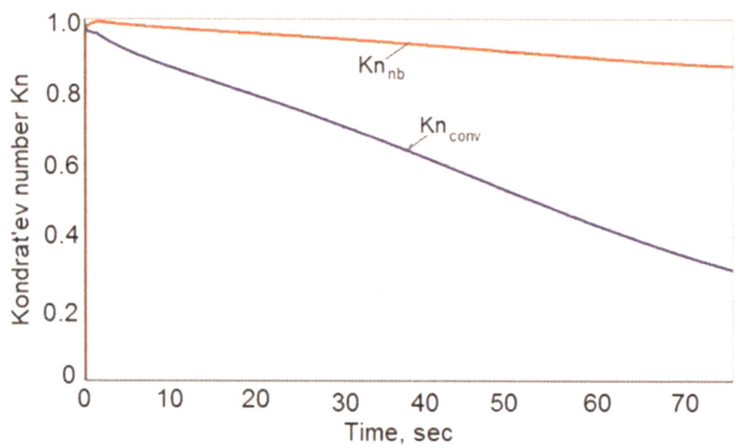

Figure 6.
Real Kn_{nb} and effective Kn_{conv} Kondrat'ev numbers versus time when quenching a 50 mm cylindrical sample in a low concentration of water salt solution at 20°C.

As one can see from **Figure 5**, the temperature difference between IQ-2 and IQ-3 processes in core cooling curve change is insignificant (see **Table 5**).

This fact opens the possibility of approximate core cooling time calculation of different steel parts using average values of Kondrta'ev numbers Kn (see **Figure 6**).

Figure 6 presents real Kn_{nb} and effective Kn_{conv} Kondrat'ev numbers as a result of the calculation of core cooling curves for a cylindrical probe 50 mm diameter quenched in still fluid with convective HTC 548 W/m^{2}K. Convective biot number for such condition is equal to 0.6 and duration of transient nucleate boiling, according to [10], is equal to 72.6 s. Kondrat'ev numbers Kn_{nb} and Kn_{conv} were calculated by the well-known equation presented in [10, 11] on the basis of solving the inverse problem for calculating HTCs as $\alpha_r = q/(T - T_s)$ and as $\alpha_{eff} = q/(T - T_m)$. Note that effective HTC is used only for core cooling time evaluation (see Eq. (5)):

$$\tau = \left[\frac{kBi_V}{2.095 + 3.867Bi_V} + \ln\frac{T_o - T_m}{T - T_m}\right] \cdot \frac{K}{aKn} \tag{5}$$

Here,

τ is cooling time in seconds; k = 1, 2, 3 for plate, cylinder accordingly; Bi_V is generalized Biot number; T_o is initial temperature; T_m is bath temperature; K Kondrat'ev form factor; a is thermal diffusivity of steel; and Kn is dimensionless Kondrat'ev number.

It is also used for cooling rate v evaluation (see Eq. (6)):

$$v = \frac{aKn}{K}(T - T_m) \tag{6}$$

According to **Figure 6**, the average effective Kn_{conv} = 0.625 while real Kn_{nb} = 0.93. It means that real generalized Biot number Bi_V = 10 [12].

There are many experimental data and databases related to effective HTCs, which are mainly used in the heat-treating industry for recipes development. That is why this issue is discussed here to be able to use cost-effective HTCs obtained by different authors. The real HTCs and real Kondrat'ev numbers are used for calculation of temperature field and residual stresses. In the last decade, it was possible due to the absence of film boiling process. The problem of elimination of film boiling process is solved effectively by the next three main approaches:

- The use of water salt solutions of optimal concentration as a quenchant where the first critical heat flux density is maximal [11].

- The use of low concentration in water of inverse solubility polymers to decrease initial heat flux density due to creation of the thin surface insulating layer (see Eq. (7)).

$$q_{in} = \frac{q_o}{\left(1 + 2\frac{\delta}{R}\frac{\lambda}{\lambda_{coat}}\right)} \tag{7}$$

Here, q_{in} is initial heat flux density of cylindrical probe covered by polymeric layer; q_o is initial heat flux density of cylindrical probe free of polymeric layer; δ is thickness of polymeric layer; R is radius of cylindrical probe; λ is thermal conductivity of steel; and λ_{coat} is thermal conductivity of insulating layer.

- The use of resonance effect generated by hydrodynamic emitters to destroy any film boiling process during quenching steel in cold fluids [13].

4. Fundamentals of transient nucleate boiling processes

When any film boiling is completely absent due to $q_{in} < q_{cr1}$, one can formulate three very important for the practice characteristics of the transient nucleate boiling process. They are:

a. For a given condition of cooling in cold fluid, the duration of establishing developed nucleate boiling is almost the same, independently of the form and size of the steel part. It can be explained by an extremely high heat exchange during shock boiling which lasts for a very short time, thereby creating a huge temperature gradient in a very thin surface layer. Core temperature during this time is not affected by shock boiling at all and cooling process at the very beginning is considered as a cooling of semi-infinity domain. It is happening because the speed of heat distribution is a finite value and during this short time of cooling, the heat does not reach layers located at the core of steel parts (it follows from Eq. 3 and Eq. 4).

b. The surface temperature of a steel part, beginning from the start of full nucleate boiling establishing, maintains at the level of boiling point of the fluid insignificantly differing from it. The so-called self-regulated thermal process is established. The initial temperature of self-regulated thermal process T_I can be evaluated as:

$$T_I = T_S + \vartheta_I, {}^oC \tag{8}$$

where

$$\vartheta_I = 0.293 \cdot \left[\frac{2\lambda(\vartheta_o - \vartheta_I)}{R}\right]^{0.3} \tag{9}$$

The temperature of self-regulated thermal process T_{II} at the end of nucleate boiling is evaluated as:

$$T_{\mathrm{II}} = T_S + \vartheta_{II}, {}^oC \tag{10}$$

where

$$\vartheta_{II} = 0.293 \cdot \left[\alpha_{conv}(\vartheta_{II} + \vartheta_{uh})\right]^{0.3} \tag{11}$$

$\vartheta_I, {}^oC$ is overheat of a boundary layer at the beginning of SRTP; ϑ_{II} is overheat of a boundary layer at the end of SRTP.

It is possible to use average temperature $(T_I + T_{II})/2$ for approximately calculating temperature fields and residual stresses during quenching of steel parts in fluids when film boiling is completely absent with the accuracy <1% .

c. As a result of long-lasting accurate experiments and appropriate analytical solutions, for the fixed initial temperatures T_o and T_m, the author [14] has formulated very important for the practice the main characteristic of the transient nucleate boiling process. It says that duration of transient nucleate boiling is directly proportional to squared thickness of steel part, inversely proportional to thermal diffusivity of material, depends on the form of steel part, and convective Biot number Bi. The generalized equation for such statement can be mathematically formulated as [14]:

$$\tau_{nb} = \Omega k_F \frac{D^2}{a} \tag{12}$$

Here, τ_{nb} can be considered as a width of noise generated by vapor bubbles, which is equal to its duration measured in seconds; Ω is dimensionless parameter depending on convective HTC; k_F is dimensionless form coefficient; D is thickness of steel part in m; and a is thermal diffusivity of steel in $m^2 s^{-1}$.

For the fixed T_o = 850°C and T_m = 20°C, the dimensionless coefficient Ω, presented by **Figure** 7, depends only on convective Biot number Bi (**Table 6**).

The heat transfer coefficients (HTCs) during transient nucleate boiling process were evaluated using Tolubinsky's equation [4]:

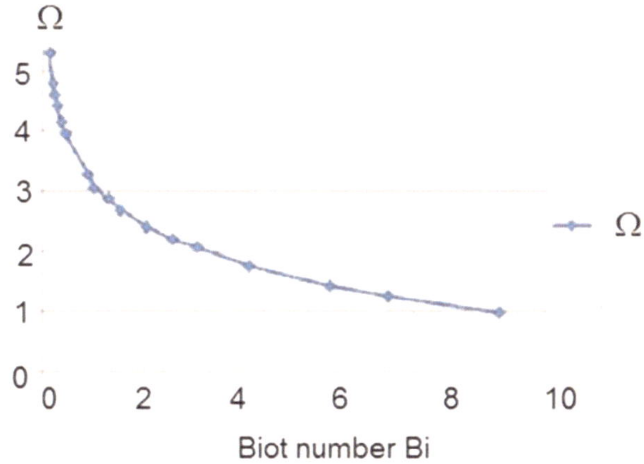

Figure 7.
Value Ω versus convective Biot number Bi [15].

Form of sample	k_F
Slab	0.1013
Cylinder	0.0432
Sphere	0.0253
Cylinder Z = D	0.0303
Cube	0.0338

Table 6.
Dimensionless form coefficients k_F for different forms of steel parts.

$$\frac{\alpha}{\lambda}\sqrt{\frac{\sigma}{g(\rho-\rho'')}} = 75\cdot\left(\frac{q}{r^{\bullet}\rho''W''}\right)^{0.7}\cdot\left(\frac{\nu}{a}\right)^{-0.2}. \tag{13}$$

According to Tolubinsky equation [4], the real HTC during nucleate boiling is calculated from the rewritten Eq. (14):

$$\alpha = 75\lambda\left[\frac{g(\rho'-\rho'')}{\sigma}\right]^{0.5}\left(\frac{a}{v}\right)^{0.2}\left(\frac{1}{r^*\rho''w''}\right)^{0.7}\cdot q^{0.7} \tag{14}$$

or

$$\alpha = cq^{0.7} \tag{15}$$

where.

$c = 75\lambda'\left[\frac{g(\rho'-\rho'')}{\sigma}\right]^{0.5}\cdot\left(\frac{a}{v}\right)^{0.2}\left(\frac{1}{r^*\rho''w''}\right)^{0.7}$; $W'' = d_0 f$.

Here, α is the real HTC during nucleate boiling process in W/m^2 K; λ is thermal conductivity of liquid in W/mK; g is gravitational acceleration in m/s^2; ρ is liquid density in kg/m^3; ρ" is vapor density in kg/m^3; q is heat flux density in W/m^2; W'' is vapor bubble growth rate in m/s; is kinematic viscosity in m^2/s; a is thermal diffusivity of liquid in m^2/s; d_o is diameter of a bubble in m; and f is frequency of a bubble departure in Hz.

Calculations of HTCs were made for maximal critical heat flux density of water salt solution which was equal to 15 MW/m^2. Dimensionless correlations of Tolubinsky and Shekriladze [4, 16] were used for evaluation of HTCs, which are presented in **Table 7**.

According to Tolubisky equation, cooling during transient nucleate boiling process is very intensive, even in still fluid if any film boiling is completely absent (see **Table 7**).

Heat flux density during initial time of quenching in cold fluids reaches almost 20 MW/m^2 when film boiling is absent (see **Figure 8a** and **b**). It is comparable with the first critical heat flux density of cold water when $q_{cr2}/q_{cr1} = 0.05$.

Heat transfer coefficients during transient nucleate boiling process can be easily evaluated if heat flux density during nucleate boiling is known (see **Figure 8a** and **b**).

According to the investigation of the author [4], during the extremely fast cooling produced by shock boiling, the first critical heat flux density q_{cr1} is very large, because the ratio (16):

$$\frac{q_{cr2}}{q_{cr1}} = 0.05 \tag{16}$$

which during conventional cooling is five times larger (see Eq. (17) published in [4, 17]:

$$\frac{q_{cr2}}{q_{cr1}} = 0.2 \tag{17}$$

According to authors [4, 17], the first critical heat flux density q_{cr1} for water at 20°C is equal to 5.9 MW/m^2 and for water at 10°C is equal to 6.5 MW/m^2. During shock boiling, the mentioned critical heat flux densities reach values 29.5 MW/m^2 and 32.5 MW/m^2 that follow from comparing Eq. (16) and Eq. (17). It means that any film boiling is completely absent during quenching in slow agitated cold water at 10–20°C. The conclusion made is in good agreement with the experimental data (see **Figures 1–3**). Since any film boiling process in many cases is completely absent, one can consider fundamental characteristics of transient nucleate boiling process as the main factors during quenching process in cold fluids. The transient nucleate boiling process exists independently of the will of people. It is happening when meteorites fall into oceans. They produce noise generated by tiny vapor bubbles. The same is happening during volcanos' activity located in oceans and near seas. Transient nucleate boiling can be:

Temperature of water solution	Tolubinsky	Shekriladze	Average
10	152,248	176,546	164,397
20	193,929	243,641	218,785
40	224,989	241,615	233,302
60	271,273	271,323	271,298

Table 7.
Real HTCs in W/m^2K during nucleate boiling process depending on the temperature of water solution when heat flux density is 15 MW/m^2.

DOI: http://dx.doi.org/10.5772/intechopen.113240

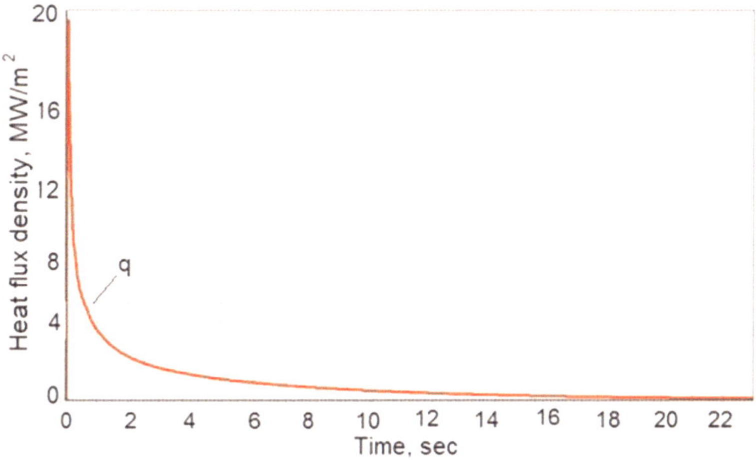

(a)

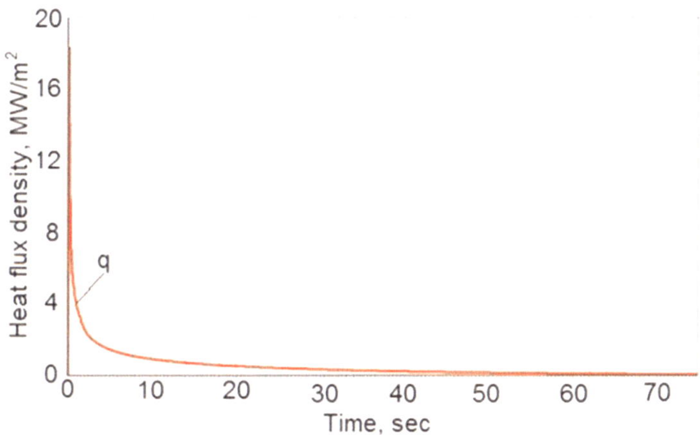

(b)

Figure 8.
Heat flux density variation versus time during quenching cylindrical specimens 25 band 50 mm in water salt solution of optimal concentration at 20°C: a – diameter 25 mm.

- Seen due to growth of bubbles and their departure.
- Heard due to noise produced by tiny bubbles.
- Smelt due to vapor that has a specific smell.
- Felt due to quench tanks' vibration during boiling.

Discovered characteristics of transient nucleate boiling process create a stable basis for:

- Making intensive quenching process into a mass production since there is no need to design costly quench tanks equipped with the powerful propellers and tanks.

- Designing absolutely new processes called austempering and martempering technological processes produced in cold water and water solutions, instead of melted salts and alkalis [18].
- Design of original intensive quenching process in water salt solutions of optimal concentration negatively charged to guarantee absence of any film boiling process during quenching [19].
- Use of resonance effect to destroy any film boiling during quenching in cold fluids [13].
- Design the method and apparatus for control quality of hardened steel parts [20].

5. Universal correlation for heating and cooling time evaluation during steel quenching

There are numerous computer codes for numerical calculation of cooling time and cooling rate of any steel part's configuration. The author of this chapter proposed a generalized universal equation for heating and cooling time evaluation while heat treating steel parts [21], which has a very simple form:

$$\tau_{eq} = \overline{E}_{eq} \frac{K}{aKn} \tag{18}$$

Here, E_{eq} is specified if value N = $(T_o - T_m)/(T - T_m)$ is known (see **Table 8**). **Table 8** provides E_{eq} value depending on N, which varies within 1.5–1000. Kondrat'ev form coefficients K (see **Table 9**) are provided by authors [11, 12, 22]. Kondrat'ev number Kn is calculated as:

$$Kn = \frac{Bi_V}{\left(Bi_V{}^2 + 1.437Bi_V + 1\right)^{0.5}} \tag{19}$$

Here

$$Bi_V = \frac{\alpha}{\lambda} K \frac{S}{V} \tag{20}$$

						$Bi_V = 0.1$					
N	**1.5**	**2**	**2.5**	**3**	**3.5**	**4**	**4.5**	**5**	**10**	**100**	**1000**
Plate	0.44	0.73	0.96	1.14	1.29	1.43	1.54	1.65	2.38	4.64	6.95
Cylinder	0.49	0.77	1.00	1.18	1.33	1.47	1.58	1.69	2.42	4.68	6.99
Sphere	0.55	0.81	1.04	1.22	1.37	1.51	1.62	1.73	2.46	4.72	7.02
						$Bi_V = 0.3$					
Plate	0.50	0.78	1.01	1.19	1.34	1.48	1.60	1.70	2.39	4.70	7.00
Cylinder	0.59	0.88	1.10	1.28	1.44	1.57	1.69	1.79	2.49	4.79	7.09
Sphere	0.68	0.97	1.19	1.37	1.53	1.66	1.78	1.88	2.58	4.88	7.18

					$Bi_V = 0.1$					
N	**1.5**	2	2.5	3	3.5	4	**4.5**	5	**10**	**100**	**1000**
$Bi_V = 0.5$											
Plate	0.53	0.817	1.04	1.22	1.38	1.51	1.63	1.73	2.43	4.73	7.03
Cylinder	0.65	0.94	1.16	1.35	1.50	1.63	1.75	1.86	2.55	4.85	7.16
Sphere	0.78	1.07	1.29	1.47	1.62	1.76	1.88	1.98	2.67	4.98	7.28
$Bi_V = 1$											
Plate	0.57	0.86	1.08	1.27	1.42	1.55	1.67	1.78	2.47	4.77	7.07
Cylinder	0.74	1.03	1.25	1.44	1.59	1.72	1.84	1.94	2.64	4.94	7.25
Sphere	0.91	1.20	1.42	1.60	1.76	1.89	2.01	2.11	2.80	5.11	7.41
$Bi_V = 2$											
Plate	0.61	0.90	1.12	1.30	1.46	1.59	1.71	1.81	2.51	4.81	7.11
Cylinder	0.81	1.1	1.32	1.50	1.66	1.79	1.91	2.02	2.71	5.01	7.33
Sphere	1.01	1.30	1.52	1.71	1.86	1.99	2.11	2.22	2.91	5.21	7.51
$Bi_V = 5$											
Plate	0.63	0.92	1.14	1.32	1.48	1.61	1.73	1.83	2.53	4.83	7.13
Cylinder	0.86	1.15	1.37	1.55	1.71	1.84	1.96	2.07	2.76	5.06	7.36
Sphere	1.10	1.38	1.61	1.80	1.94	2.08	2.20	2.30	3.00	5.29	7.58
$Bi_V = \infty$											
Plate	0.64	0.93	1.15	1.33	1.49	1.62	1.74	1.84	2.54	4.84	7.15
Cylinder	0.87	1.16	1.38	1.56	1.72	1.85	1.97	2.08	2.77	5.07	7.38
Sphere	1.11	1.39	1.62	1.80	1.95	2.09	2.20	2.31	3.00	5.30	7.60

Table 8.
Coefficients E_{eq} depending on dimensionless value N that varies from 1.5 to 1000, taking into account different values of generalized Biot numbers Bi_V [19].

Form of sample	**Kondrat'ev form factor K**
Slab	$K = \frac{L^2}{9.87}$
Cylinder	$K = \frac{L^2}{5.783}$
Sphere	$K = \frac{R^2}{9.87}$
Cylinder Z = D	$K = \frac{R^2}{8.2505}$
Cube	$K = \frac{L^2}{29.61}$

Table 9.
Kondrat'ev form factor K for different forms of steel parts.

Example 1: Cylindrical sample 50 mm diameter, made of AISI (American Iron and steel institute) 1040 steel, is quenched from 860°C in low concentration of still water salt solution at 20°C [23, 24]. Maximal surface compression residual stress reaches its maximal value at the moment when core temperature of cylinder reaches 430°C. Calculate cooling time from 860–430°C to provide further self-tempering process of surface layers and obtain high surface compression residual stress. For the given process, N = (860–100°C)/(430–100°C) = 2.3. According to **Table** 7, for Bi_V = 10 and N = 2.3, E_{eq} = 1.36. Kondrat'ev form factor K = R^2/5.783 = 108.1 × 10^{-6} m^2. Average value of thermal diffusivity of AISI 1040 steel a = 5.4 × 10^{-6} m^2/s. According to Eq. (18), τ = (1.3 × 108.1 × 10^{-6} m^2)/(5.4 × 10^{-6} m^2/s × 0.93) = 29.3 s. Approximately τ= 30 s.

Example 2: Cylindrical forging 50 mm at its end, made of AISI 1040 steel, has a temperature 950°C. It is quenched intensively in spray water salt solution of low concentration creating condition when Bi_V tends to infinity. Of still water salt solution at 20°C. Calculate cooling time from 970–480°C to provide further self-tempering process of surface layers and obtain high surface compression residual stress and improved mechanical properties due to high temperature thermomechanical treatment. For given condition, N = (970–20°C)/(480–20°C) = 2. According to **Table** 7, for Bi_V equal infinity and N = 2, E_{eq} = 1.16. Kondrat'ev form factor K = R^2/5.783 = 108.1 × 10^{-6} m^2. Average value of thermal diffusivity of AISI 1040 steel for interval temperatures from 500–950°C is a = 5.6 × 10^{-6} m^2/s. According to Eq. (18), τ = (1.16 × 108.1 × 10^{-6} m^2)/(5.6 × 10^{-6} m^2/s × 1) = 22.4 s.

Forging requires a shorter cooling time because it is sensitive to crack formation due to overheating.

6. Conclusions

1. For given condition of cooling, the time of transient nucleate boiling establishment is approximately the same independently of the form and size of the quenched steel part. It is explained by an extremely forced heat transfer process which can be considered as a cooling of semi-infinity domain.

2. Transient nucleate boiling is considered as a self-regulated thermal process when surface temperature maintained at the level of boiling point of a fluid insignificantly differs from its boiling point. In this case, the real surface temperature can be replaced by its average value.

3. Length of transient nucleate boiling process is directly proportional to squared thickness of steel part, inversely proportional to thermal diffusivity of material, and depends directly on the form of steel part and convective Biot number Bi if initial temperatures T_o and T_m are fixed.

4. The formulated principles are the basis for design of new technologies and their accurate control. The mentioned principles provide correct temperature fields' calculation of quenched steel parts without performing, in many cases, painstaking costly experiments.

DOI: http://dx.doi.org/10.5772/intechopen.113240

Author details

Nikolai Kobasko
Intensive Technologies Ltd., Kyiv, Ukraine

*Address all correspondence to: nkobasko@gmail.com

References

[1] Tensi HM. Wetting kinematics. In: A Handbook "Theory and Technology of Quenching". Berlin: Springer Verlag; 1992. pp. 93-116

[2] Totten GE, Bates CE, Clinton NA. Handbook of Quenchants and Quenching Technology. Materials Park, USA: ASM International; 1993. p. 513

[3] French HJ. The Quenching of Steels. Cleveland, Ohio, USA: American Society for Steel Treating; 1930

[4] Tolubinsky VI. Heat Transfer at Boiling. Kyiv: Naukova Dumka; 1980. p. 316

[5] Lykov AV. Theory of Heat Conductivity. Moscow: Vysshaya Shkola; 1967. p. 600

[6] Guseynov S, Buikis A, Kobasko N. Mathematical statement of the task with the hyperbolic equation of the thermal conductivity for the intensive quenching of steel and its solution. In: Proceedings of 7th International Conference "Equipment and Technologies for Heat Treatment of Metals and Alloys". Vol. II. Kharkov: KhTI; 2006. pp. 22-27

[7] Kobasko NI. Uniform and intense cooling during hardening steel in low concentration of water polymer solutions. American Journal of Modern Physics. 2019;**8**(6):76-85

[8] Kobasko NI, Aronov MA, Ichitani K, Nasegava M, Noguchi K. High compressive residual stresses in through hardened steel parts as a function of Biot number. In: Resent Advances in Fluid Mechanics, Heat &Mass Transfer and Biology. Harvard: WSEAS Press; 2012. pp. 36-40

[9] Kobasko NI. High Quality Steel Vs Surface Polymeric Layer Formed during Quenching. Germany: Lambert Academic Publishing; 2019. p. 98

[10] Kobasko NI. Basics of quench process hardening of powder materials and irons in liquid media. European Journal of Applied Physics. 2022a;**4**(3): 30-37

[11] Kobasko NI. Regular thermal process and Kondrat'ev form factors. In a book "Intensive Quenching Systems: Engineering and Design". USA: ASTM International; 2010. pp. 91-106

[12] Kondrat'ev GM. Thermal Measurements. Moscow: Mashgiz; 1957

[13] Kobasko NI. UA Patent No. 109572. Intensive hardening method for metal components. 2013

[14] Kobasko NI. Transient nucleate boiling as a law of nature and a basis for designing of IQ technologies. In: Proc of the 7th IASME/WSEAS International Conference on Heat Transfer, Thermal Engineering and Environment. Moscow: WSEAS; 2009. pp. 67-76

[15] Kobasko NI, Aronov MA, Powell JA. Intensive Quenching Processes, Quenchants and Quenching Technology. Vol. 4F. ASM International, USA: ASM Handbook; 2023

[16] Shekriladze IG. Boiling heat transfer: An overview of longstanding and new challenges. In: In a book "Film and Nucleate Boiling Processes". USA: ASTM International; 2010. pp. 229-284

[17] Kutateladze SS. Fundamentals of Heat Transfer. New York: Academic Press; 1963

[18] Kobasko NI. Unexpected forced heat exchange phenomenon to be widely used for new quenching technologies development. Physics Letters. 2021;**9**(8): 179-191

[19] Kobasko NI. Self-regulated thermal process taking place during hardening of materials and its practical use. Theoretical Physics Letters. 2022;**10**(2): 273-287

[20] Kobasko NI, Moskalenko AA, Dobryvechir VV. UA Patent No. 119230. Method and apparatus for control quality of hardened in fluids metal components. 2019

[21] Kobasko NI. Thermal equilibrium and universal correlation for its heating – Cooling time evaluation. Theoretical Physics Letters. 2021;**9**(10):205-219

[22] Kondrat'ev GM. Regular Thermal Mode. Moscow: Gostekhizdat; 1954

[23] Kobasko NI, Moskalenko AA, Logvynenko PN, Dobryvechir VV. New direction in liquid quenching media development. Thermophysics and Thermal Power Engineering. 2019;**41**(3): 33-40

[24] Kobasko NI. Mechanism of film boiling elimination when intensively quenching steel parts in water polymer solutions of low concentration. Global Journal of Science Frontier Research: A (Physics and Space Science). 2020;**20**(7): 39-56

Chapter 3

A Brief Review of Techniques of Thermal Enhancement in Tubes

Shamoon Jamshed

Abstract

Heat transfer enhancement in tubes is not a novel idea. These tubes are used in several engineering devices like heat exchangers, boilers, evaporators, refrigerators, and air conditioners, to name a few. To date, these tubes are undergoing an evolutionary period, since engineers are trying new ways to improve the heat transfer (or enhance the heat transfer). The main cause is the pressure loss that occurs due to friction and the limitation of the surface area of the tube. The passive techniques to overcome this loss are more common due to cost-effectiveness. Thus, common passive techniques include grooves inside the tube surface, grooves on both the inner and outer surface, or putting inserts within the tube. Modern techniques are utilizing nanofluids, that carry nano-materials inside the heat transfer fluid to enhance heat transfer. To quantitatively gauge the heat transfer enhancement, the heat transfer effectiveness is computed. This chapter deals with the study of the above-mentioned techniques in some detail and discusses minimizing entropy generation rate in groove tube(s). Also, a bird's-eye view of the nanofluids and their usage for heat transfer enhancement has been seen.

Keywords: heat transfer, groove tube, entropy generation, nanofluids, pressure loss, friction, effectiveness, nanofluids

1. Introduction

Heat exchangers (HX)s are quite frequently used in industry. There are various types of heat exchangers available and their use depends upon the particular requirement. The most common type of heat exchanger is the indirect contact type where the fluids exchange heat employing a surface or surfaces lying between the two fluids. Thus, the name indirect applies. In indirect-contact type HXs, there are further classifications, such as tubular type, extended surface type, and plate-type heat exchangers. The extended type heat exchangers have been the subject of much research and this chapter focuses on techniques of optimization of these tube-type HX with reference to mechanical optimization in particular.

The work on tube-type heat transfer has been of much interest for about 50 years. After the late 1970s, the buzzing of global warming due to Chloro Floro Carbon (CFCs) and carbon footprints, engineers and scientists started to improve the devices that are majorly responsible for expelling carbon into the atmosphere. Mostly, among

IntechOpen

others such as gasoline and human waste, refrigeration devices were found to be responsible for emitting CFCs at a much higher rate. Several international conferences on heat transfer and symposia led the engineers to overcome this problem multifold by improving the device itself rather than the refrigerant. Hence, many research works were presented, attributed, and led to the design of tubes that have extended surfaces.

The works of Bergles, Ralph Webb, and Naryanamuruthy are highly recognized in the community. Arthur E. Bergles is a very well-known personality in the field of augmented heat transfer. He has written many papers and authored many books on enhanced heat transfer (**Figure 1**).

Ralph Webb was a Professor Emeritus of Mechanical Engineering at Penn State University. He wrote the book, Principles of Enhanced Heat Transfer. He performed research by discussing various heat transfer phenomena occurring in heat exchangers, and the design optimization of heat exchangers. He was a veteran consultant in more than 100 major companies and wrote 275 papers in international journals. He died in 2011.

Adrian Bejan is also a professional in the heat transfer field. He has done exacerbated work on entropy generation minimization. This topic has not been touched by any other for heat transfer surfaces. Adrian Bejan has written books on entropy minimization in heat exchange devices. He has also done work on a newer topic called Constructural Law but this is not in the scope of this book or chapter.

2. Methods of heat transfer enhancement

There are many ways of heat transfer enhancement. Mainly, the classification is based on the mode of contact with the fluid and the surface. Two main classes are active and passive. Active methods involve the use of some external power or resource to generate heating. This energy will definitely increase the heat transfer rate. Nevertheless, the surface vibration or the fluid vibration is also used in active means of heat transfer enhancement. The electrostatic field is another technique for heat transfer. Jet impingement is frequently used for heat and cooling as well. It is used for cooling gas-turbine blades where the cold fluid from various holes along the blade cross section is impinged on the surface, to make the blade surface cooler. This type of cooling is called film cooling. Film cooling is also used for cooling liquid rocket engines where

Figure 1.
Pioneers of the field of enhanced heat transfer.

the cold fuel gets heated thereby cooling down the engine's heat and thus making the combustion effective.

The passive method involves direct heat transfer by modification in the hardware. By hardware, we mean the surface modification. This can be done by roughening the surface. By manufacturing, a good metal finish can be achieved. But the roughness, in fact, improves the heat transfer. A roughened surface increases the surface area and therefore improves the heat transfer rate. On the other hand, a poor surface finish should not be to such an extent to decrease material quality. The tube surface should also be capable of bearing high temperatures and pressure. If cost is of less issue, another method is the extended surface. This requires adding the fins on the outer or inner surface of the tube. Sometimes, the fins are welded on the outer surface. External surfaces are used if external heating is required. For inner surfaces, since welding a metal piece is not possible on the inner side, grooves are made. This is also sometimes done by making a rectangular sheet with inclined grooves etched into the sheet at a certain included angle (usually 45 or more). Then, the sheet is rolled and welded into forming a tube. Usually, the non-grooved part is now acting as an extended surface but on the inner side. The tubes used in air-conditioning condensers consist of 10 to 15 mm in diameter. This small diameter thus makes the grooving quite difficult. But, due to modern machining techniques (or the rolling technique as mentioned above), these groove-making processes are quite common now. The groove depth or the fin height on the inner side is classified as micro since $h/D < 0.03$ [1].

Sometimes the tube is itself rotated in the form of a coil. This also improves heat transfer. It does not necessarily increase the heat transfer area but (depending on the pumping power) the fluid now has heat stored in it for a long time, since the travel path is reduced due to coiling. This makes the heat transfer process effective. Coiled exchangers also offer not only a higher heat-transfer coefficient but also a more effective use of pressure drop due to large curvatures. Thus, the designs are used in cases where a large pressure drop is required besides the heating advantage. The coil design allows the management of high temperatures and extreme temperature differentials without high-induced stresses or costly expansion joints. High-pressure transmission capacity with the ability to fully clean the service-fluid flow area also adds a good characteristic to this type of design.

Adding additives nowadays is a very common technique. Additives are added in the form of nano-particles that increase the heat transfer properties of the mixture. **Table 1** shows these different modes of heat transfer enhancement.

The different types of geometrical formats for enhancing heat transfer are pictorially shown in **Figures 2** and **3**.

3. Literature on enhanced tubes

3.1 Important terminologies

There have been several works on enhanced heat transfer. The reason for this immense amount of research on this topic is obvious, that is the tubes with enhanced surfaces have proven to be a successful device in achieving the goal. Since nothing is perfect, there are methods and technologies still under discussion and debate to make tube-side heat transfer effective. Four terms are widely used in this regard.

Passive	Active (when some external power is applied)
Rough surface	Mechanical aids (stirring, rotating surface)
Extended surface	Surface vibration
Swirl flow device	Fluid vibration (1 Hz to ultrasonic)
Coiled tubes	Electrostatic field (DC or AC)
Additives (in the form of particles)	Impingement of jet (normal or oblique)
Hybrid	
A roughened surface with twisting tape insert	

Table 1.
Different techniques of heat transfer enhancement.

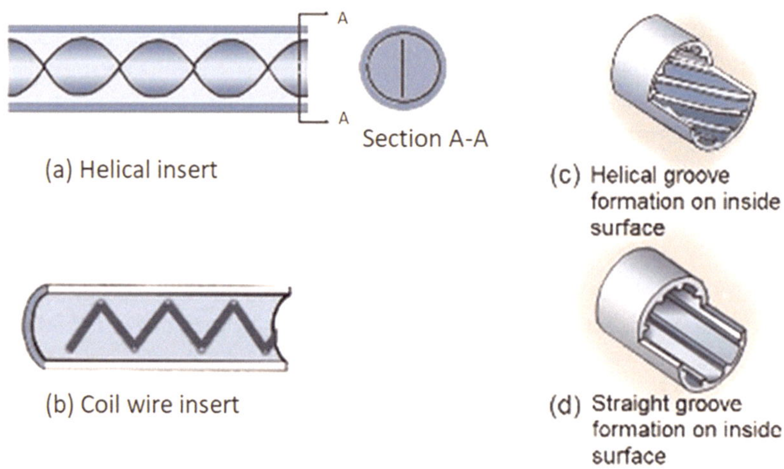

Figure 2.
Tube designs showing groove tube and insert in a tube.

- The friction factor
- The Nusselt number
- The Thermal enhancement factor

The entropy generation rate.

3.1.1 Friction factor

The friction factor is the dimensionless pressure drop occurring along the length of the tube. Thus, it's directly related to the pumping power. The more the pumping power is, the lesser the pressure drop will be.

3.1.2 Nusselt number

Nusselt number is defined as the dimensionless heat transfer coefficient. It is usually given as.

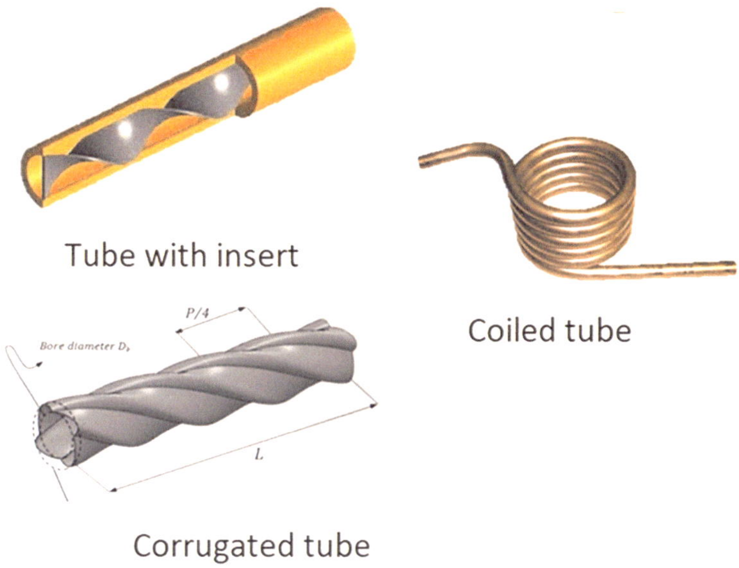

Figure 3.
Different types of active heat transfer techniques.

$$Nu = hD/\lambda. \tag{1}$$

Where 'h' is the heat transfer coefficient, D is the hydraulic diameter of the tube, and λ is the thermal conductivity of the heat transfer fluid. In the analysis of the heat transfer enhancement, the Nu and f are computed. These are computed for a plain tube (with no enhanced surface) and with an enhanced surface.

3.1.3 Thermal enhancement factor

There is another term defined as thermal enhancement factor or thermal effectiveness which is given as,

$$\eta = (Nu/Nu_o)/(f/f_o). \tag{2}$$

Where the subscript 'o' is for the plain tube. This term η is used for characterizing the thermal enhancement capacity of a tube. It should be >1, for obvious reasons, but if it is not, then it means that the friction factor term is dominating. This means that the tube surface is not effective in enhancing heat and pumping power is consumed in increasing the pressure to overcome the frictional loss. It will also depend upon the design that how much η can be afforded, but it is clear that it is required that it should be >1.

3.1.4 Entropy generation rate

With this in view, a fourth term is needed to compute the overall heat transfer loss/gain. This (in a way) is a missing piece. This term is entropy generation rate. This is given as

$$\dot{S}_{gen} = \frac{q\Delta T}{T^2} + \frac{\dot{m}\Delta p}{T\rho} \tag{3}$$

Where, $\dot{S}_{gen}$ is the rate of entropy generation; q is the heat flux; ΔT is the temperature difference across the tube ends; $\dot{m}$ is the mass flow rate; Δp is the pressure drop along the tube length. T is the bulk temperature and ρ is the fluid density.

3.2 Experimental studies

Most of the work on the enhanced-surface tube-heat transfer studies is experimental. There are notable works by Webb, Bergles, and Kakaç as referenced in [1, 2].

The work that has been done on enhanced-surface-tube research during the last century is mostly experimental-based. In these studies, there is a vast literature available revealing the notable works of Webb, Bergles, Sunden, and Sadik Kakaç, in the works of Webb et al. [1] and Kakaç [2]. Naryanamuruthy and Webb [1] discussed the heat transfer in helical groove tubes. He discussed the performance of tubes with respect to the Colburn j-factor ($StPr^{2/3}$) and friction factor. It was concluded that the tubes (when compared with tubes with roughened surfaces) were equally good in depicting the heat transfer enhancement behavior. Jensen [3] discussed the fin dimensions, i.e., height and depth on the surface of the tube. He discussed in detail the effect of fin dimensions on the friction factor and Nusselt number. It was noticed that in the micro-fin tubes, the friction factor experiences a long delay for the flow to become fully turbulent. The flow becomes fully turbulent at Re = 20,000. Chen et al. [4] conducted a study on corrugated tubes. These tubes are not finned or grooved, rather they have a surface turned in the form of a wavy shape. Four-start (the number of grooves/fins/wavy shape seen at the start of the tube) tube was studied in detail. Results were compared with the study of Ravigururajan and Bergles [5], and Srinivasan and Christensen [6] This study concluded that, as far as the trend is concerned, Chen's results are not in agreement with Srinivasan's. Their trend is similar to that of Ravigururajan for heat transfer with a difference of 3000 W/m^2K. Another researcher K. Gregory [7], investigated eight helical finned tubes and one smooth tube. Results of the heat transfer coefficient and friction factor were obtained and compared with the work of Webb et al. [8]. Results were comparable, in a good estimate, and within the prediction errors from 30 to 40%. Aroonrat et al. [8] determined the Nusselt number and the friction factor for the Reynolds number range of 4000 to 10,000. The tube with the largest helix angle predicted the highest Nusselt number. The helix angle in the tubes was varied from 0 to 90° with the highest Nu obtained in the case of 60° with 0.5-inch pitch. The same trend was followed for the friction factor.

3.3 Numerical studies

Finned geometries were also examined from the numerical simulation perspective. Due to the advent of modern computers, performing complex flow simulations is no more a challenge. Therefore, there has been an increasing trend in Computational Fluid dynamics (CFD) based simulations over the past two decades. A study by Liu and Jensen [9] shows the work on several fin profile formations within a tube. Liu et al. [10] found that there is seldom variation in the fin profile if they were of rectangular or triangular shape. While the round profile under-predicted the friction factor and Nusselt number which was figuratively 7–10% less than the rectangular-fin geometry. Kim et al. [11], utilized a Finite Element Method (FEM) based solver for predicting the flow and heat transfer effects and validated their results. Jasinski [11] and [12] used the CFX code. He determined the heat transfer and friction parameters for different helical angles in micro-finned tubes. This numerical study was a bit different from other researchers in the way that Jasinski determined the entropy generation rate in the tubes. The minimum entropy generation rate was found. The minimum rate of entropy ($\dot{S}_{gen}$) was found for different helix angles, and it was predicted to be low for 700 tubes at Re = 60,000. Pirbastami et al. [13] predicted the results of Aroonrat et al. [8] using numerical simulations. Pirbastami modeled the complete three-dimensional geometry of the tube. However, the chapter did not take into account the solid region. This, in my view, is not correct. The chapter merely mentioned (by taking the point of Aroonrat) that the flux on the tube's inner wall was reduced to 30% from what was applied on the top outer wall. This approach, in the author's opinion, is not correct even though the results luckily matched the experimental data. This could be due to the reason that the wall was only 2 mm thick in comparison to the tube length, hence, flux variation did not affect the net results. Nevertheless, this study was a good approach in terms of validation but performed with many assumptions. H. R Kim et al. [14] and S. Kim et al. [14, 15] studied the fluid flow and flux in a twisted elliptical tube. Research work was similar to the works of Piotr, Jensen, and Pirbastami, as reported. This study, though, contained in-depth work since the effects of friction factor, Colburn j-actor, volume, and area were studied in this chapter.

Jamshed et al. [16] did an in-depth study on Aroonrat's geometry and also analyzed customized geometries of groove tubes. They studied different tubes with different pitch and helix angles. It was found that the tube with the lowest pitch gives better performance but at a bit lower Reynolds number of 5000. Also, the entropy generation rate is minimum with the tube with the lowest pitch.

4. Trends in the tube-side heat transfer research

If we monitor the trends over the last two decades, most of the work has been done numerically. **Figure 4** shows the trends that have been occurring in the last 60–70 years. This shows that significant studies have been done in the numerical domain. Now, the work is more inclined toward entropy minimization and optimization of Heat Transfer Fluid (HTF) with techniques like nanofluids. It should be noted that the scale of progress is just schematic and does not replicate the actual number of publications.

5. An example of the validation case

A simple case was selected for validating the results of Aroonrat (through CFD) as mentioned in the literature review above. Jamshed et al. [16] have a detailed study on it. Here are some glimpses of the work done. This is mentioned to make the audience familiar with the validation process. The tubes selected were already described by Aroonrat [8]. The length of the tube was 2 m and the material was Steel. Thermocouples were mounted for temperature measurement on the tube. Constant heat flux condition was maintained at the wall with magnitude 3500 W/m^2. The groove geometry is shown in **Figure 5**.

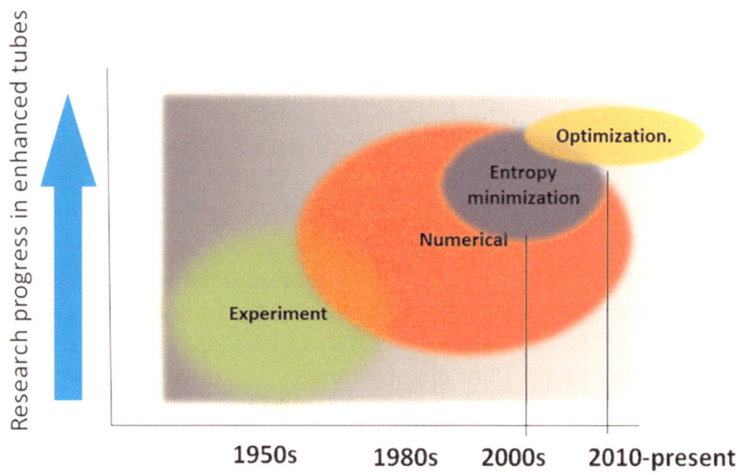

Figure 4.
Progress in tube-side heat transfer enhancement research.

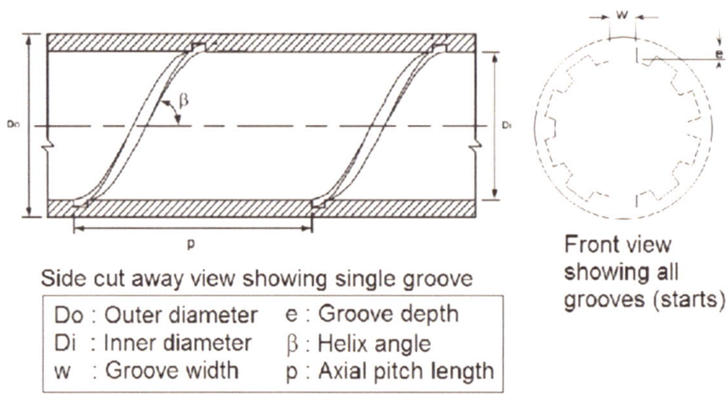

Figure 5.
Cutaway of the groove tube, from [9].

The modeling was done in the software Gridgen v 15. This included the geometric modeling of the tube and the meshing of the fluid+solid region. Later, for the CFD analysis, ANSYS Fluent v16 was used. The turbulence model used was k-ω SST with default pressure-based solvers. Details are well explained in the papers by Jamshed et al. [16]. Due to the symmetric nature of the helical grooves and the flow, the modeling was simplified using the technique of Jensen [3]. This technique implied the modeling of 1/n the geometry where n is the number of grooves (start), then circumferentially modeling the segment from one groove to the other. After that, this geometry was extruded (lengthwise) over a single helical pitch length. The model is considered to be symmetric about the tube axis. Thus, this helically extruded geometry will be repeated until the tube length of 2 m is achieved. The geometrical shape of the tube that was modeled and meshed is shown in **Figure 6**. Flux is applied on the top wall with the solid portion shown in red color.

Other boundary conditions include the input of the velocity inlet on one side of the tube and the pressure outlet on the other side. Periodic symmetry was applied to the cyan-colored region shown in the **Figure 6**. It should be noted that this symmetry boundary condition is of the rotational type where the rotational axis lies on the z-axis.

5.1 Friction factor

The friction factor is computed from CFD simulations. The results for different tubes are plotted in **Figure 7**. These results are for GT_02, GT_04, GT_06, and GT_08.

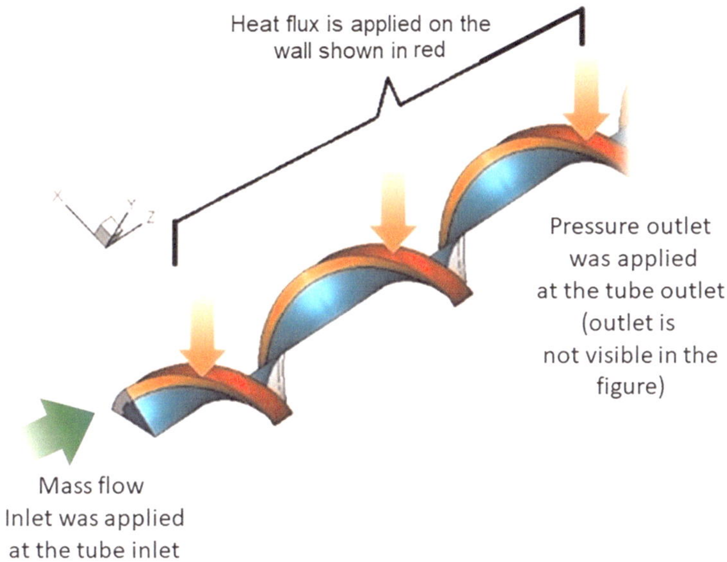

Figure 6.
Modeled tube geometry with the flux applied on the top of the wall.

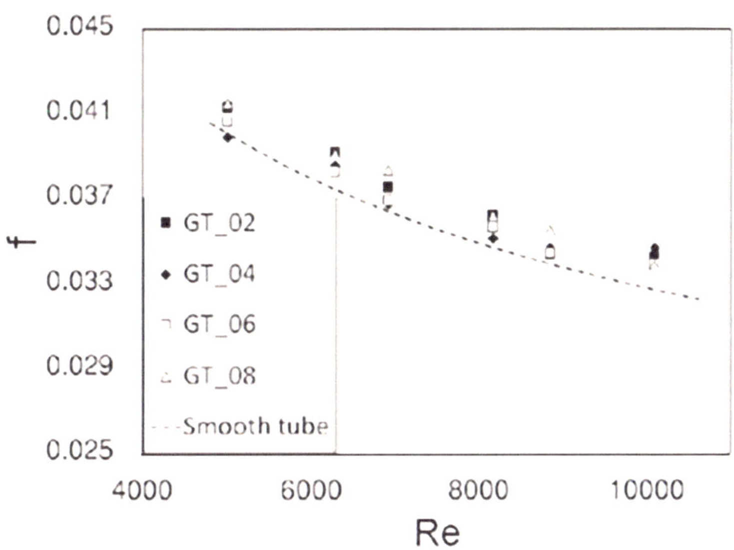

Figure 7.
Effect of friction as a function of Reynolds number on the performance of different groove tubes.

The numbers in the nomenclature indicate the pitch value in inches. These values indicate that GT_02 has the highest value in terms of pitch length. It is also indicated that all the tubes have a higher friction factor than the smooth tube which is expected. The trend is also logical, observing a decrease in value as a function of the Reynolds number. The Reynolds number is computed based on the hydraulic diameter of the smooth tube. GT_02 being the highest is also logical since the smallest pitch length tube should reveal a greater resistance to fluid motion and significant pressure drop.

5.2 The Nusselt number

The Nusselt number is computed based on the formula mentioned in Eq. (1) above. The heat transfer coefficient is computed as

$$\boldsymbol{h}_{avg} = \frac{q''}{T_{avg,w} - T_{avg,f}} \tag{4}$$

Where $T_{avg,w}$ is the average wall temperature and $T_{avg,f}$ is the bulk fluid temperature obtained by averaging the inlet and outlet temperatures. This havg is used in the formula of Nu. **Figure 8** shows the behavior of the Nu as a function of the Reynolds number and brings the GT_02 again in the lead. The increasing trend is due to the same physics as followed for the friction factor. However, the high friction also increases turbulence which increases the heat transfer rate. Since the flux is constant, and as the Reynolds number gets increased, the Nusselt number increases much faster. A high Reynolds number makes the boundary layer thin. This has a rapid temperature change from Twall to Tbulk and thus the difference minimizes. This increases h_{avg}, and thence the Nu. This increasing trend will be much more pronounced for very high Reynolds numbers, but here it is not as such due to lower values of Reynolds numbers. These trends can be seen in the paper of Piotr Jasinski [17].

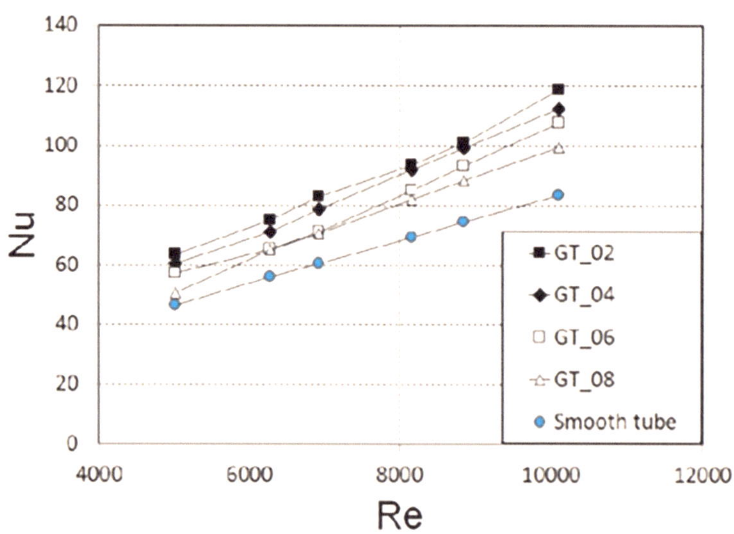

Figure 8.
Effect of Nusselt number as a function of Reynolds number on the performance of different groove tubes.

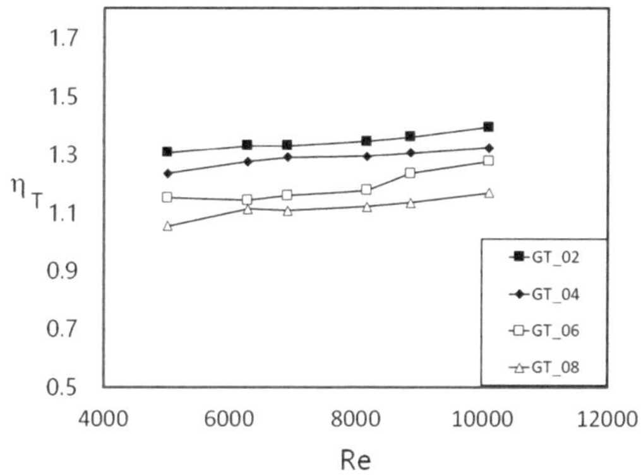

Figure 9.
Thermal enhancement factor for different tubes.

5.3 The thermal enhancement factor

As described above, the Thermal Enhancement Factor (TEF) or (η) is a very important parameter in checking the performance of the enhanced surfaces. This parameter is the ratio of the Nusselt number ratio to the friction factor ratio. Each ratio is the ratio of the enhanced surface/tube to the plain surface/tube. **Figure 9** shows the TEF of the tubes discussed above. This graph shows that the tube with 2-inch pitch is at the lead. This is also expected since both the Nusselt and the friction factor were above the mark in comparison with the other tubes and the smooth tube. Overall, it can be concluded that the tube performance can be easily examined through numerical simulations. TEF can also be predicted by first computing the Nusselt number and the friction factor. The complete performance matrix will be established by determining the Entropy Generation rate which will be discussed next.

5.4 Entropy generation minimization

Bejan [18] coined the concept of entropy generation minimization. It is not that the idea did not exist earlier. It existed there, but the concept was that an engineering system (or a device) working on the first principle, i.e., the thermodynamic principles, their performance must be evaluated on these principles as well. Earlier, before the works of Adrian, the problem of heat transfer enhancement was evaluated on the principle of the first law of thermodynamics, i.e., the energy is conserved. The problem was not seen from the viewpoint of the second law of thermodynamics. In terms of entropy, the second law states that entropy never decreases and it keeps on increasing. If you notice the title of this paragraph, it is written as entropy generation minimization. It never said, entropy minimization. Therefore, the system (enhanced tube) is discussed based on the way to minimize the generation rate of entropy. After all, a system with losses generates more entropy so in a way we are decreasing the losses (or trying to minimize the losses) thus increasing the availability as mentioned by Bejan in his texts [19, 20].

In exchangers, that is our focus, there is a trade-off between heat transfer irreversibility and friction work irreversibility. This can be easily understood by describing the ratio of entropy ratio of the augmented tube to the entropy generated due to the plain tube [17, 21]. In respect of the enhanced tube, the numerical works done by Jasinski [11], and Bharat Kumar et al. [21] are notable. Works related to entropy by Majeed [22] and Zheng et al. [23] are on the axial groove tube.

Adrian Bejan mentions that the entropy generation ratio can be used to establish a design point for an augmented tube. A curve can be plotted showing the entropy ratio ($S_R = \frac{\dot{S}_{gen}}{\dot{S}_{gen,o}}$) as a function of the Reynolds number. In this curve, there must be a point at which the irreversibilities due to the friction of the fluid and the irreversibilities due to heat intersect. Therefore, a generalized curve can be generated. The right side of the intersection point shows the friction effect domination and the left side shows the thermal effects domination (**Figure 10**).

For the study of helical grooved tubes that are mentioned above, the same analogy was applied and the curve for SR was plotted. The entropy ratio curve depicts the optimized point for a tube giving the whole picture at various Reynolds numbers. The ratio is a function Reynolds number that looks like a v-shaped curve. This curve has a minimum point that lies at SR = 1. This shows that this is the point where the heat transfer effect and frictional losses have become equal. This curve is, in fact, a combo of two curves as shown in **Figure 11**.

5.4.1 Entropy generation minimization-practical example

After some years of the publication of the above work, a very similar work was published. The work of Bharat et al. [21] was on the same theme of entropy minimization in internally made helical finned tubes. His finding on the design optimization using entropy-based work resembles the research work of Jamshed et al. [16, 24], whereas their work on the multi-objective optimization of the thermal and pressure-based entropy generation is similar to the paper by Jamshed [25] on multidisciplinary optimization.

In the work of Bharat et al. [21], the entropy generation rate study was the entropy generation rate study was accompanied by the Pareto-optimal design technique. In

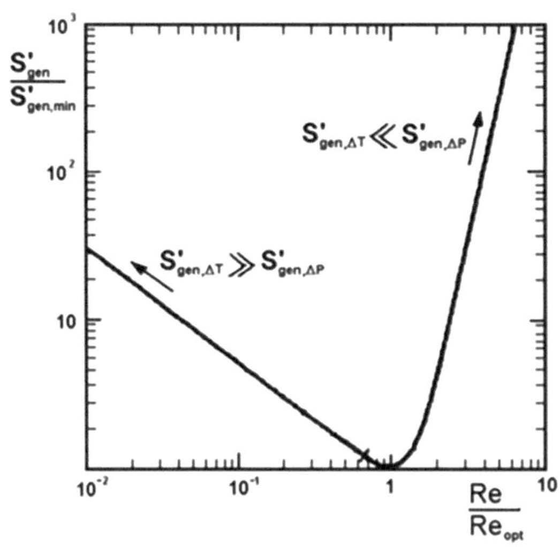

Figure 10.
Entropy generation ratio versus Reynolds number curve for a tube, Bejan [18].

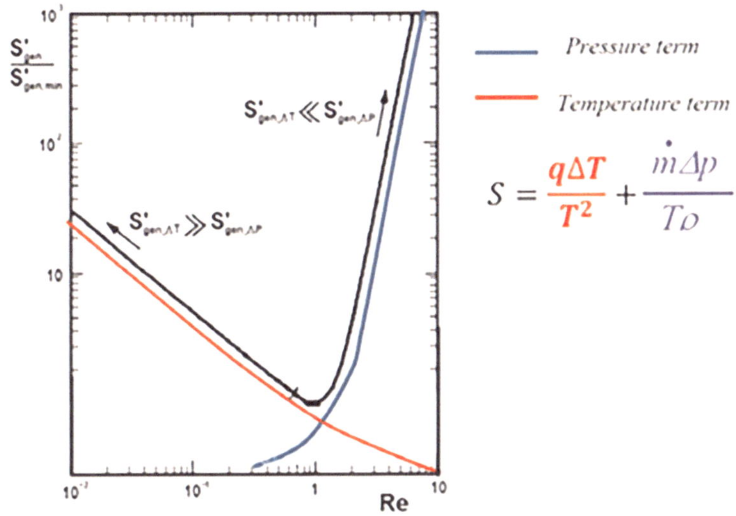

Figure 11.
Entropy ratio curve showing the effects of the pressure term and the temperature term.

this technique, the height and pitch of the helical fin were chosen as design variables. The optimal solution (i.e. entropy rate minmization) is compared to the solution of multi-objective problem. This considers either minimization of pressure or the heat, but not both. However, the author also claimed that reducing pressure and heat transfer terms individually, may not be the right matrix to measure the performance of entropy. This is because entropy due to both terms is a strong function of the Reynolds number. Engineers can explore the optimal charts as per their design needs. In the study, Bharat et al. concluded that the entropy due to the pressure term is optimized at 25% of the heat transfer entropy generation. But this performance can be substantially different for other Reynolds numbers. This is shown in **Figure 12**.

Similar work was done using multi-disciplinary optimization of the grooved tubes by Jamshed et al. [25]. The Design of Experiments (DoE) technique was used with a D-optimal design taking the groove depth, helix angle, and the number of grooves as variables. The effect of the entropy generation rate was obtained at all of the design points. It was found that the tube with 1-inch of pitch length gave the maximum entropy. Now the optimum Reynolds number was computed, based on the same analogy described above, that the intersection of heat and pressure term will give the optimum Reynolds number. Thus, it was found to be 5000 for the 1-inch pitch tube as shown in **Figure 13**. A correlation was developed for the tube showing $S_R = f(Re)$. This tube shows that the optimal point lies where $S_R = 1$ and it lies at the optimum Re of 5000. The entropy ratio curve and the correlation are shown for the tube with Ns = 30, e = 0.5 mm, and p_L = 25.4 mm (1-inch).

6. Nanofluids—a new avenue in the heat transfer enhancement domain

Nanofluids are fluids with nanoparticles. As the name indicates, nanoparticles are very minute-sized particles of the order of 10^{-9} m. Nanotechnology-based techniques

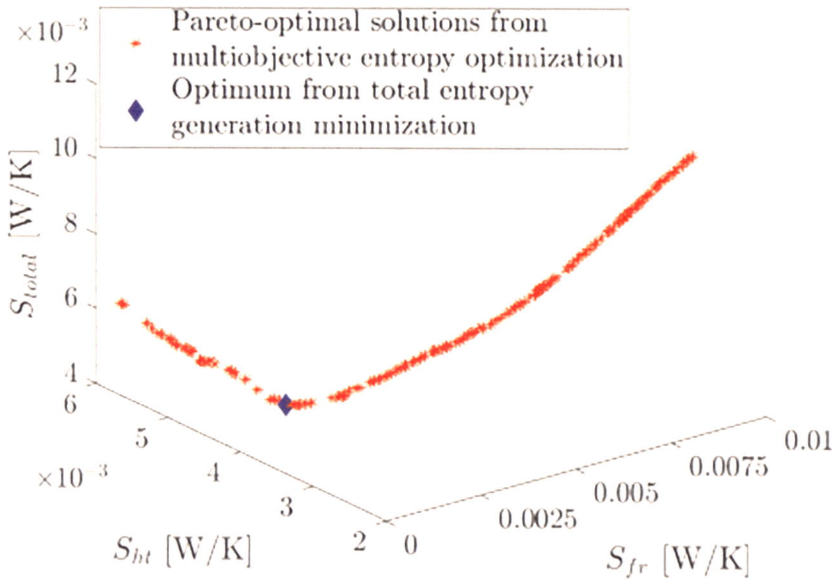

Figure 12.
Entropy generation rate minimization through Pareto-optimization, from Bharat et al. [21].

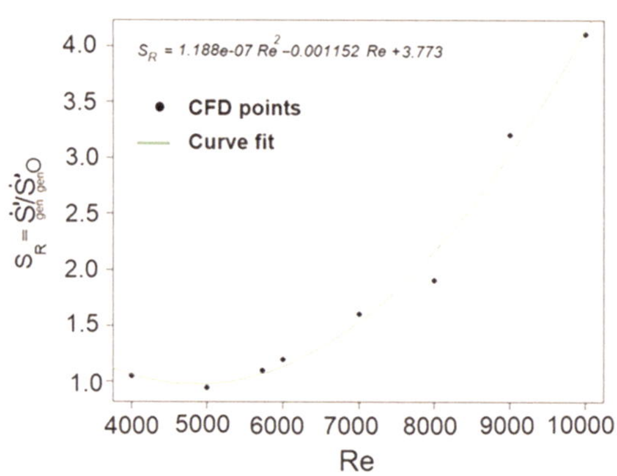

Figure 13.
The entropy ratio curve and the correlation shown for the tube with groove numbers = 30, groove depth = 0.5 mm, and pitch length = 25.4 mm (1-inch).

could be used to produce these particles. These particles when added to a fluid of lower thermal conductivity, increase the heat transfer characteristics of the medium.

As mentioned by Mahmoud Salem [26], the nanofluid is a fluid in which nano-sized particles are suspended in a base fluid. The solute thus forms a colloidal solution of particles in the base fluid. These particles are made up of metals, carbon nano-tubes, or oxides of metals. The base fluid is generally water, ethylene glycol, or oil. Research has been quite embryonic in understanding nano-particle physics thoroughly, but these particles have been found to possess great properties in enhancing heat transfer. Thus, these nanofluids have made a place in applications like engines, domestic refrigerators, chillers, and even in flue gas temperature-reduction of a boiler.

6.1 Nanofluids for heat exchangers

The thermal performance of a heat exchanger (HX) is greatly affected by several factors. As discussed above, the design or geometry is one of the factors that had been seen thoroughly. Fluid properties and the type of flow (turbulent or laminar) also affect the HX performance a lot. Since fluid properties have a major contribution in enhancing the performance of an HX, nanofluids second this thought. **Figure 14** shows the enhancement ways of thermal performance of an HX.

The effects of density, particle size, and concentration have been greatly discussed in a paper by Stephan et al. [27]. Nanoparticles, on the one hand, increase the heat transfer coefficient (HTC) but also increase the pressure drop. It was also found that the shape also matters in enhancing heat transfer such as the increase in thermal conductivity. Pressure drop is affected by the shape of the particles where it was found lower in the case of the spherically shaped particles. Stephan et al. studied the effect of TiO_2 on the water and computed the result using mathematical modeling. It was found that all three properties, i.e., thermal conductivity, density, and viscosity were increased by 1.98, 2.61, and 2.03% respectively compared to the base fluid. With the constant flow rate, the heat transfer coefficient was also increased by 1.25% while the pressure drop was also increased by 2.03%.

The effect of flow rate with a constant concentration of 0.4% was also observed. In the variable flow rate, condition it was found that the HTC increased by 17.20%, and the pressure drop was found to be increased by 60%. This means that increasing the flow rate increases the heat transfer by a large magnitude but also increases the pressure drop. Thus, pumping power is increased. However, changing concentration, does not increase the HTC a lot, but can affect the overall performance of the heat exchangers since they need to run for longer durations. A small percentage amount can give larger volumes of benefits at a later time.

Stephan concluded and recommended that nanoparticles of high conductivity and small size may be selected. However, the concentration of the particles must be adjusted with the volumetric flow rate to elude huge pressure drops.

Thong Le Ba et al. [28] conducted a study of nanofluids in a circular tube. They have used CFD for the analysis of particles. The effect of SiO_2-P_{25} particles in the solvent water/ethylene glycol was observed in a circular tube. Volume concentrations were 0.5, 1.0, and 1.5%. The Reynolds number in the flow was from 5000 to 17 k.

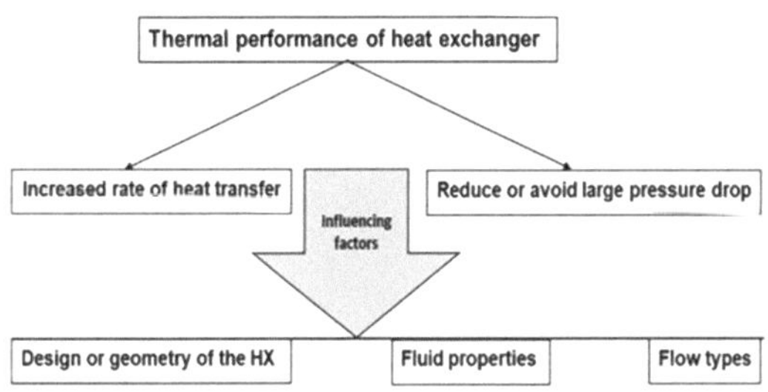

Figure 14.
Influencing factors on the thermal performance of heat exchanger, from [26].

While the heat flux was 7955 W/m^2. A constant heat flux condition was maintained. In the other case, a constant wall-temperature of 340.15 K was separately provided on the walls of the tube and results were obtained. The turbulent flow in the tube was monitored and it was concluded that constant heat flux gave similar results as that of a temperature-dependent case with lower cost. Nusselt number and pressure drop were increased with the increasing concentration and the flow rate of the nanofluid. This was also in good agreement with the experimental results.

With this literature, it can be concluded that nanofluids are an effective means of heat transfer enhancement. Proper effectiveness will be achieved through skills and dexterity in achieving the right proportions of concentration amount of the particles, their conductivity, and density, and this can be easily pre-analyzed through simulations. As the nanofluids technology is expensive, it is also difficult to conduct repeated experiments to see the effect of concentration, density, or thermal conductivity.

7. Conclusion

Passive techniques mostly increase the surface area at the expense of the pumping power (which increases due to friction of the increased surface area). Thus, common passive techniques include grooves inside the tube surface, grooves on both the inner and outer surface, or putting inserts within the tube. First, in this chapter, mainly, the passive technique, such as groove, to enhance heat transfer is seen through a numerical study. CFD technique has been employed for validation of the friction factor results and the Nusselt number from published literature. The results were quite in good agreement with the experimental data. Secondly, the entropy generation rate minimization is also seen in these tubes which is a less-seen avenue of computing heat transfer enhancement in groove tube(s). It has been observed that a cut-off value of the pressure loss and the heat gain due to entropy describes the optimum point of the performance. Hence the optimum Reynolds number is defined. And lastly, a bird's-eye view of the nanofluids and their usage for heat transfer enhancement has been seen since modern techniques (for heat transfer increase) are utilizing nanofluids, that carry nano-materials inside the heat transfer fluid to enhance heat transfer. To quantitatively gauge the heat transfer enhancement, the heat transfer effectiveness is computed in all the cases.

Author details

Shamoon Jamshed
Institute of Space Technology, Karachi Campus, Karachi, Pakistan

*Address all correspondence to: shamoonjamshed@gmail.com

References

[1] Narayanamuruthy, Webb R. Principles of Enhanced Heat Transfer. 2nd ed. Taylor & Francis; 2005. p. 217. ISBN 9781591690146

[2] Kakaç S, Bergles AE, Mayinger F, Yüncü H. Heat Transfer Enhancement of Heat Exchangers. Springer; 1999

[3] Jensen MK, Vlakancic A. Experimental investigation of turbulent heat transfer and fluid flow in internally finned tubes. International Journal of Heat and Mass Transfer. 1999;**42**: 1343-1351

[4] Chen XD, Xu XY, Nguang SK, Bergles AE. Characterization of the effect of corrugation angles on hydrodynamic and heat transfer performance of four-start spiral tubes. Journal of Heat Transfer. 2001;**123** (December):1149-1158

[5] Ravigururajan TS, Bergles AE. General correlations for pressure drop and heat transfer for single-phase turbulent flow in internally ribbed tubes. Augmentation of Heat Transfer in Energy Systems. 1985;**52**:9-20

[6] Srinivasan V, Christensen RN. Experimental investigation of heat transfer and pressure drop characteristic of flow through spirally fluted tubes. Experiments in Thermal Fluid Sciences. 1992;**5**:820-827

[7] Gregory JZ, Louay MC, Pedro JM. Experimental determination of heat transfer and friction in helically-finned tubes. Experiments in Thermal Fluid Sciences. 2008;**32**:761-775

[8] Webb R, Narayanamurthy R, Thors P. Heat transfer and friction characteristics of internal helical-rib roughness. Journal of Heat Transfer. 2000;**122**:130-142

[9] Aroonrat K, Jumpholkul C, Leelaprachakul R, Dalkilic AS, Mahian O, Wongwises S. Heat transfer and single-phase flow in internally grooved tubes. International Communications in Heat and Mass Transfer. 2013;**42**:62-68

[10] Liu X, Jensen MK. Geometry effects on turbulent flow and heat transfer in internally finned tubes. Journal of Heat Transfer. 2001;**123**:1035-1044

[11] Kim JH, Jansen KE, Jensen MK. Analysis of heat transfer characteristics in internally finned tubes. Numerical Heat Transfer, Part A, Applications. 2004;**46**:21

[12] Jasinski P. Numerical optimization of flow-heat ducts with helical micro-fins, using entropy generation minimization (EGM) method. In: Recent Advances in Fluid Mechanics and Heat & Mass Transfer. Florence, Italy: WSEAS Press; 2011. pp. 47-54

[13] Pirbastami S, Moujaes SF, Mol SG. Computational fluid dynamics simulation of heat enhancement in internally helical grooved tubes. International Communications in Heat and Mass Transfer. 2016;**73**:25-32

[14] Kim HR, Kim S, Kim M, Park SH, Min JK, Ha MY. Numerical study of fluid flow and convective heat transfer characteristics in a twisted elliptic tube. Journal of Mechanical Science and Technology. 2016;**30**(2):719-732

[15] Kim S, Kim M, Park YG, Min JK, Ha MY. Numerical investigation of fluid flow and heat transfer characteristics in a

helically-finned tube. Journal of Mechanical Science and Technology. 2017;**31**(7):3271-3284

[16] Jamshed S, Qureshi S, Khalid S. Numerical flow analysis and heat transfer in smooth and grooved tubes, Wessex transactions on engineering sciences. In: 11th International Conference on Advances in Fluid Mechanics. Vol. 105. UK: Wessex Institute of Technology; 2016. p. 283

[17] Jasinski P. Numerical study of the thermo-hydraulic characteristics In a circular tube with ball turbulators. Part 2, heat transfer. International Journal of Heat and Mass Transfer. 2014;**74**: 473-483

[18] Bejan A. Entropy Generation through Heat and Fluid Flow. USA: Wiley; 1982

[19] Bejan A. Advanced Engineering Thermodynamics. USA: Wiley; 1998

[20] Bejan A. Entropy generation minimization, the new thermodynamics of finite-size devices and finite-time processes. Journal of Applied Physics. 1996;**79**(3):1191-1217

[21] Pidaparthi B, Li P, Missoum S. Entropy-based optimization for heat transfer enhancement in tubes with helical fins. Journal of Heat Transfer. 2021;**144**(1):012001

[22] Ramadhan AA, Al Anii YT, Shareef AJ. Groove geometry effects on turbulent heat transfer and fluid flow. Heat and Mass Transfer. 2013;**49**:185-195

[23] Zheng N, Liuand P, Shan F, Liu Z, Liu W. Turbulent flow and heat transfer enhancement in a heat exchanger tube fitted with novel discrete inclined grooves. International Journal of Thermal Sciences. 2017;**111**:289-300

[24] Jamshed S, Qureshi S, Shah A, Hussain A. Investigation of entropy generation rate and its minimization in helical grooved tubes. Journal of Mechanical Science and Technology. 2018;**32**(10):4983-4991. Available from: www.springerlink.com/content/1738-494x(Print)/1976-3824. DOI: 10.1007/s12206-018-0946-6

[25] Jamshed S. Multidisciplinary optimization in helical grooved tubes for heat transfer enhancement. In: Engineering Science & Technology. UK: Universal Wiser Publisher; 2021

[26] Ahmed MS. Nanofluid: New Fluids by Nanotechnology. London, UK: IntechOpen; 2019. DOI: 10.5772/intechopen.86784

[27] Louis SP, Ushak S, Milian Y, Nems M, Nems A. Application of nanofluids in improving the performance of double-pipe heat exchangers—A critical review. Materials. 2022;**15**:6879. DOI: 10.3390/ma15196879

[28] Le Ba T, Gróf G, Odhiambo VO, Wongwises S, Szilágyi IM. A CFD study on heat transfer performance of SiO_2-TiO_2 Nanofluids under turbulent flow. Nanomaterials. 2022;**12**:299. DOI: 10.3390/nano12030299

Section 2

Heat Transfer – Advanced Applications

Chapter 4

Advanced Shell-and-Tube Longitudinal Flow Technology for Improved Performances in the Process Industry

Marcello Garavaglia, Fabio Grisoni, Marta Mantegazza and Marco Rottoli

Abstract

Several heat exchanger technologies have been developed in the second half of the former century thenceforth for addressing a multiplicity of incumbent topics shaping the discussion in the technical community and the economics of the process industry. In the frame of shell-and-tube layout, longitudinal flow deserves a peculiar place. Initially conceived for addressing requisition for reduced vibration and fouling accumulation and later recommended in case of limited allowable pressure drops, it proved valuable in replacing segmental layouts whereas weight and footprint come into the picture and reliability matters. Structural increases in the cost of raw materials and expectation for extended operational continuity push the industry in the direction of more efficient and dependable technologies. This chapter focuses on the EMbaffle® design, among the most reputed longitudinal flow shell-and-tube technologies, whose extensive adoption in oil and gas, chemical sector, and renewable power generation in the last two decades allows some fair yet not exhaustive considerations. After a concise introduction to the feature of longitudinal flow technology and to EMbaffle® basic design equations, measures of performance will be discussed. Comparison with conventional technologies will be outlined. Selected realizations will be critically presented and their potential for effective market penetration duly assessed.

Keywords: shell-and-tube heat exchanger, enhanced heat transfer performance, CO_2 reduction, improved continuity of service in heat transfer lines, augmented reliability of process equipment, advancements in process equipment technology

1. Introduction

In the second half of the twentieth century, the impressive growth of the process industry, triggered by broad economic development, promoted the emergence of new technologies. Process engineers started thinking in terms of innovative designs capable of addressing issues whose frequency could not be neglected in view of demanding plant performances.

Shell & tube (S&T) heat exchangers did not make an exception. As the workhorse in the industry, they provided different services under diverse temperature and pressure operating conditions and fluids, emerging as a reliable solution where reliability and performance meet. Double and triple segmental layouts and TEMA types (i.e. G, H, J and X), designed for addressing lower pressure drop requirements by saving at the same time the advantages of the robust and rugged S&T layout, added to well-established single segmental layouts based on E and F TEMA types [1–3]. While standard design proved robust and reliable and guaranteed safe operations in critical conditions and even harsh environments for decades, weak performances could occur with negative impacts on maintenance costs and exchanger life; yet consequences in terms of line shutdown due to equipment failure could hardly be underestimated. Novel designs aimed to overcome limitations of conventional designs (i.e., segmental), the ones that suffered with crude oil and high gravity residues are often rich in impurities, sometimes of a sticky nature; yet larger flow rates driven by increased production levels, solicited inception of vibrations which could damage the tube bundles. Moving from the existing design characterized by baffles placed normally to the direction of the flow and inducing cross shell-side flow, new baffles were conceived and tested in a variety of services, founding their specific field of application at the end.

The case of No Tube in Window (NTIW) embeds a peripheral shell-side chamber to streamline the cross flow toward a quasi-longitudinal flow and found wide application, whereas large flow rates require minimization of cross components of velocity impacting on the bundle. Since the tubes in the peripheral chamber cannot be properly supported, they would be most susceptible of vibration, and therefore are suppressed, such as all the tubes pass through all the baffles (**Figure 1**).

Disk & Doughnut, by replacing double segmental baffle with rounded baffles shaped as alternate disks and annulus, increases the longitudinal versus transversal contribution of the flow reducing the pressure drops. So, they are still widely used, whereas pressure drops are a limiting factor (**Figure 2**). Twisted tubes® work on tube

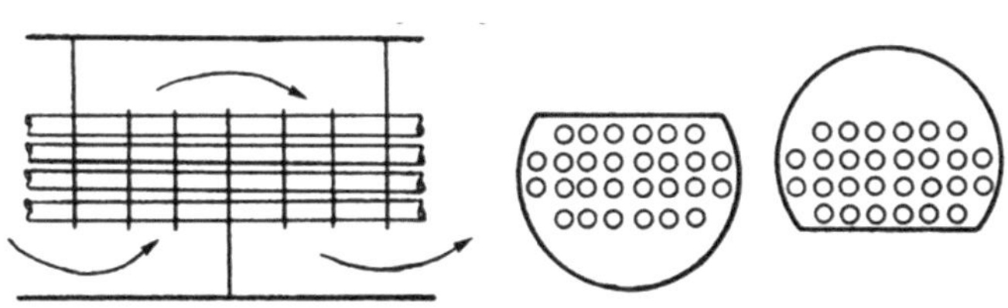

Figure 1.
NTIWs layout.

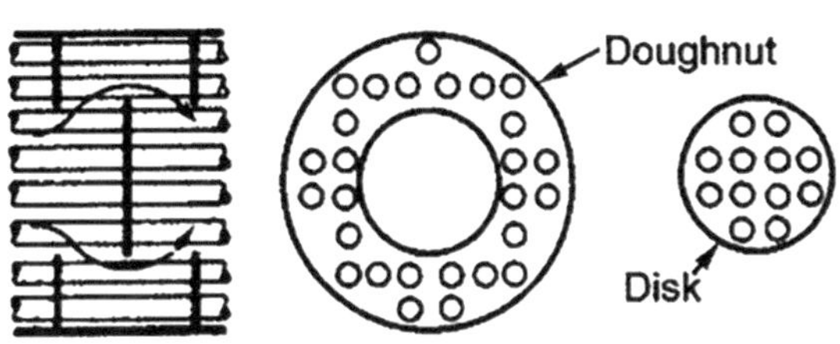

Figure 2.
Disk & Doughnut layout.

geometry by "twisting" the tubes around their axis, therefore inducing a quasi-longitudinal shell-side flow under turbulent profile. This baffles-free layout is profitably used whereas fouled fluids dictate low accumulation to avoid of deposits on the heat transfer surfaces [4]. Finally, Helixchanger® emerged as a disruptive design for addressing heavy fouled fluids thanks to inclined segmental baffle geometry which favors processing of the same under reduced deposition. It was proved in several services featuring viscous and dense fluids [5].

The above solutions were developed as improved conventional design within the frame of the conservative oil and gas world.

In the 1970s, a team of engineers of Phillips Petroleum Co., while investigating the damage that occurred to a waste heat recovery unit due to induced fluid-dynamics vibrations, suggested changing the existing baffle layout and moving to a radically new one. By replacing the same with rods, initially arranged over elliptical-shaped disks and later over circular shapes, as it became customary, the Rod Baffle design (**Figure 3**) was on the air [6, 7]. The initial aim had been to find a promising solution for vibration issues but, due to the pure longitudinal flow layout which clears vibration at its root, it emerged for addressing a multiplicity of topics, becoming a standard in the industry, especially in the North American world.

Rod Baffle triggered new thinking that eventually resulted in the flourishing of designs in recent decades. Replacement of rods with strips and with structures capable of increasing the confinement of the tubes just demonstrated the vitality of the original concept, and here is where EMbaffle® originated.

EMbaffle® is based on an expanded metal grid type of baffle, fully supporting the tubes and improving heat transfer while containing pressure drops (**Figure 4**). Conceived and initially developed by Shell Corporation at the outset of the new millennium for addressing the fouling issues experienced in the Group Refineries, the technology, owing to the continuous developments by Brembana&Rolle (B&R), found promising fields of application due to benefits of longitudinal design jointed with patented open structure feature. The next chapter will explore the benefits and wide applicability of longitudinal flow based on peculiar technical features [8–10].

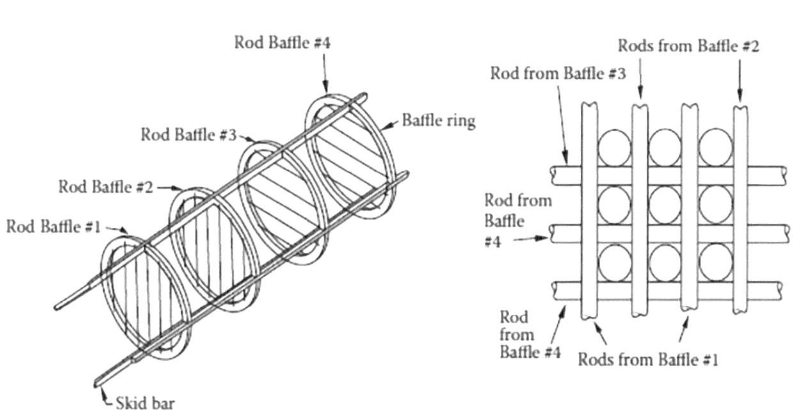

Figure 3.
Rod Baffle layout.

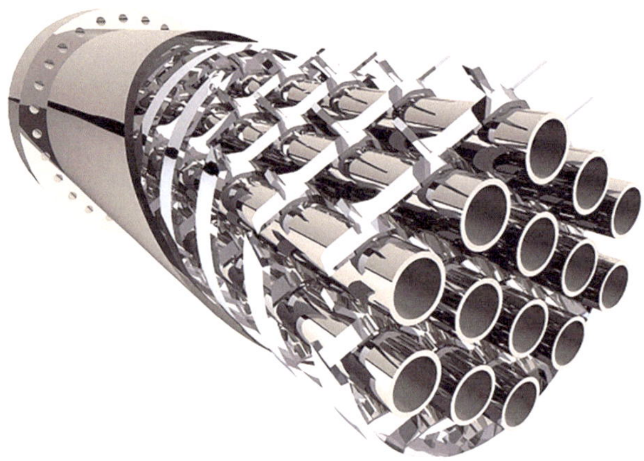

Figure 4.
EMBaffle® layout.

2. Longitudinal-type heat exchanger design

Longitudinal flow layout, i.e., establishment of parallel flow between shell-side and tube-side fluids, allows pure co/counter-current relative flow to be assessed and results in a minimum temperature approach (augmenting the efficiency of the unit); it promotes the reduction of required heat transfer area, footprint, and weight. Also, parallel flow allows avoidance of cross-components of the velocity of shell-side fluid on the tube bundle, inception of vibrations, extended maintenance frequency, and improved unit reliability.

Yet, the less tortuous and shorter path followed by the shell-side fluid with respect to cross flow configuration produces lower pressure drops which favor routing of larger mass flow rates, improving the performance of existing units in plant modernization projects (extending the operating windows). The above is at the basis of lower fouling accumulation in longitudinal flow-based designs, being suppressed by any hurdle, which would cause low velocity and recirculation regions where sticky fouling might grab and adhere.

It must be remarked that longitudinal flow is not by itself likely to trigger effective heat transfer unless promotion mechanisms are put in place; here is where the different technologies differ and where the proprietary concept generally lies.

Bundle construction follows the consolidated conventional baffles knowledge, simply replacing the same with Rod style baffles or other engineered solutions.

Although the operation of the units does not pose specific issues, maintenance may greatly be advantaged either in terms of reduced frequency of maintenance or occurrence of bottlenecking; cleaning of the bundle is generally achieved with the usual high-pressure water-jet methodology.

Rod Baffle paved the way for the deployment of several longitudinal flow designs, all sharing the above benefits.

The open geometry of tube supporting elements fits for large flow rates. Moreover. installing a vapor belt, i.e., a peripheral ring with slots, downstream the inlet nozzle through homogenization of the flow ensures prompt establishment of longitudinal flow.

Thinking that Rod Baffle has been primarily conceived for supporting the tube bundle in order to prevent vibration-induced tube damage, it's no surprise that the performance in terms of heat transfer was not focused on by the designer, further missing the advantage of the low pressure drops attainment; so Rod Baffle resulted to be the equipment of choice specifically for addressing vibration issues.

So the potential of longitudinal flow had still to be exploited well beyond Rod Baffle; the ensuing deployments will be addressed in the next chapters.

The above considerations permit us to sketch the applicability frame of the longitudinal flow layouts. Firstly, the process engineer will consider their adoption in all the cases' efficiency, effectiveness, or operating window, as well as large mass flow rates are involved. Secondly, by aggregating the technical features, it is possible to deduct the fields of application quite straightforwardly: large flow rates under low pressure drops along longitudinal flow suggest applicability to a gas service, while longitudinal flow jointly with low-temperature approach encompasses a generalized feed-effluent service; longitudinal flow under contained pressure drop is typical for a fouled service and so forth. The list includes conventional and emerging gas services, from gas pre/inter/after cooling in compression stations to CO_2 utilization in advanced power cycles, from gas dehydration in offshore platforms to synthesis of hydrogen, in the fertilizing industry along the nitrogen production chain and in more general gas–gas interchanger applications.

3. EMbaffle® technology

Diamond-shaped grid characterizes the geometry of EMbaffle® design and governs its thermo-hydraulic behavior.

Patented know-how based on so-called expanded metal production process, allows the design of different grid shapes for any tube diameter of practical interest.

Fluid dynamics of flow across a grid structure has been well known since the 1990s [11–13]; experimental studies conducted in laminar and turbulent regimes describe the grid as a turbulence promoter by (i) destroying the laminar regime and (ii) superimposing additional turbulent modes to the existing flow structure. The three-dimensional nature of the added modes can be expressed in terms of a single parameter, named turbulence intensity, which shows a prominent peak in the proximity of the grid. While a simplified theoretical formulation of such parameter, which allows its correlation with the geometry of the grid is still not available, CFD tools allow it to be represented conveniently as a function of related Reynolds and Prandtl numbers.

Fluid-dynamics is more complex whereas tubes are inserted into the diamond-shaped grid; flow is pushed inside the grid by the driving pressure force and expands downstream. Turbulent modes generated during the expansion of the fluid stream after the grid add to the modes generated in no-tube configuration; expansion cooperates in increasing turbulence through the well-known expansion-cone effect. This depends on the angle the grid is inclined, and it achieves a maximum in correspondence with a narrow range of slopes.

Proprietary design tools, supported by CFD and experimental validation process, permit to qualify each single grid geometry in terms of intensity of turbulence for specified Reynolds number. Being heat transfer coefficient and pressure drops directly related to the turbulent intensity, it follows that, for assigned process conditions, the EMbaffle® exchanger may be customized in terms of required duty over allowable pressure drops by selecting the grid shape and its number for a specified tube diameter (**Figure 5**).

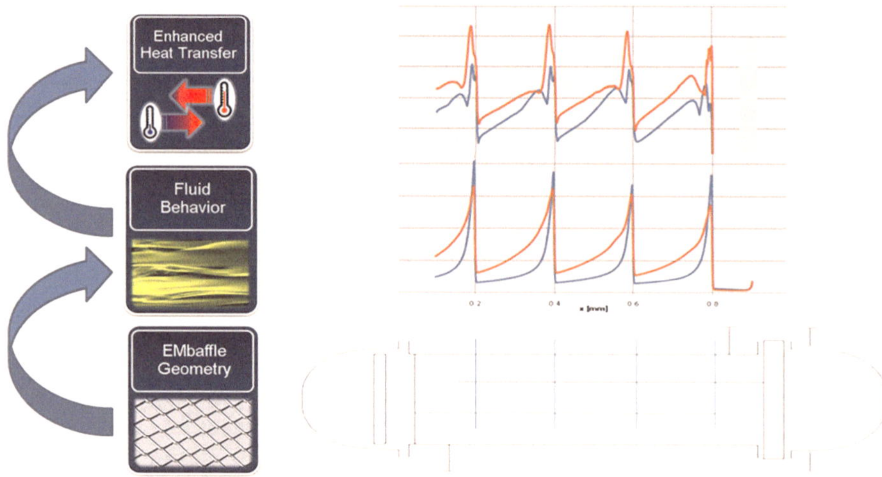

Figure 5.
Turbulence versus Duty. Governing turbulence for customizing heat transfer in EMbaffle® (illustration rights by B&R).

The ability to shape the geometry of the grid for increasing the shell-side heat transfer and accommodating the desired number of tubes adds a valuable degree of optimization. Yet the open structure of the grid geometry plays a role in preventing fouling accumulation typical of segmental baffles, which ultimately explains the reasons of the early fortune of the technology.

The University of Cambridge [14] developed the SmartPM tool for evaluating fouling deposition in S&T heat exchangers, actually integrated in the HTRI platform [15]; EMbaffle® resulted in very low fouling accumulation over several Refinery services.

In the early 2000s, a major effort was put into play in order to identify the governing equations for EMbaffle® technology. Moving from Rod Baffle's established equations for shell-side heat transfer coefficient and pressure drops, experimental tests, conducted in recognized third-party test bench facilities, permitted to finalize the grid-dependent coefficients for applicable thermo-hydraulic equations, either for laminar and turbulent regimes, in the Re range 1×10^2 to 1×10^5.

Heat transfer correlations for laminar and turbulent flow are respectively:

$$\mathrm{Nu} = C_{\mathrm{L}}\,\mathrm{Re}_{\mathrm{h}}{}^{0.6}\mathrm{Pr}^{0.4}\left(\frac{\mu_{\mathrm{b}}}{\mu_{\mathrm{w}}}\right)^{0.14} \tag{1}$$

$$\mathrm{Nu} = C_{\mathrm{T}}\,\mathrm{Re}_{\mathrm{h}}{}^{0.8}\mathrm{Pr}^{0.4}\left(\frac{\mu_{\mathrm{b}}}{\mu_{\mathrm{w}}}\right)^{0.14} \tag{2}$$

The geometry coefficient functions, C_{L} and C_{T}, account for the enhancement due to the cross flow at the shell entrance and exit and by-pass flow. The Reynolds number is calculated as:

$$\mathrm{Re}_{\mathrm{h}} = \frac{\rho U_{\mathrm{S}} D_{\mathrm{h}}}{\mu_{\mathrm{b}}} \tag{3}$$

where U_{S} is the shell-side velocity and D_{h} is the thermal characteristic diameter.

The shell-side velocity is calculated with the continuity equation, using the following expression for the shell-side flow area:

$$A_s = \frac{\pi}{4}\left(D_s^2 - N_T D_T^2\right) \tag{4}$$

The thermal characteristic diameter is calculated as:

$$D_h = \frac{4\left(P_T - \frac{\pi}{4} D_T^2\right)}{\pi D_T} \tag{5}$$

Experimental measures permitted to validate the general heat transfer correlations and the coefficient C_L and C_T for different grid geometries. In **Figure 6** the measured Nusselt number as a function of Reynolds number is reported. The kink in the prediction curve sets the transition between laminar and turbulent regimes.

Pressure drops are calculated as the sum of the longitudinal flow component and the baffle flow component:

$$\Delta P = \Delta P_L + \Delta P_B \tag{6}$$

The expression for the longitudinal component is:

$$\Delta P_L = \frac{2\rho f_F L_i U_S^2}{D_P} \tag{7}$$

where D_P is the hydraulic characteristic diameter, f_F the Fanning friction factor and L_T the length of the tubes.

D_P is calculated as follows:

$$D_P = \frac{4\left[\frac{\pi}{4}\left(D_s^2 - N_T D_T^2\right)\right]}{\pi\left(D_s + N_T D_T\right)} \tag{8}$$

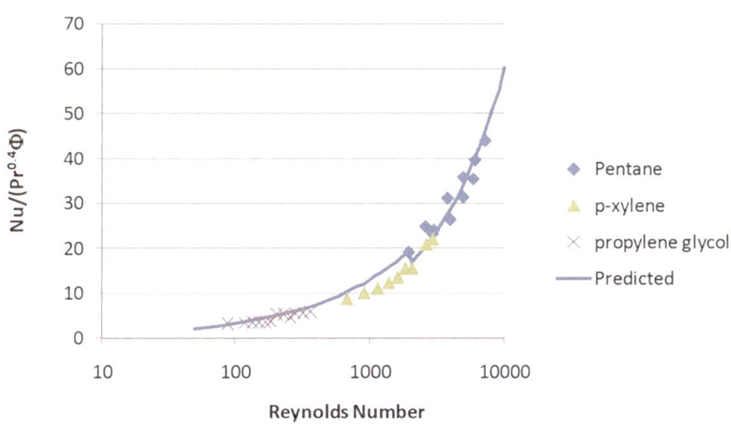

Figure 6.
Measured Nusselt number as a function of Reynolds number.

The friction factor is based on the following expression:

$$f_F = \begin{cases} \dfrac{16}{\mathrm{Re}_p}, & \text{laminar } \mathrm{Re}_p \\[2ex] \dfrac{0.079}{\mathrm{Re}_p^{\,0.25}}, & \text{turbulent } \mathrm{Re}_p \end{cases} \tag{9}$$

The baffle pressure drop is calculated using the baffle velocity U_B and a baffle loss coefficient K_B:

$$\Delta P_B = K_B N_B \frac{\rho U_B^2}{2} \tag{10}$$

where N_B is the number of the baffles.

The baffle velocity is determined using the continuity equation with the following definition of the baffle flow area:

$$A_B = A_S - A_R - A_{EM} \tag{11}$$

A_R is the ring area, while A_{EM} is the projected area of the EMbaffle® grid upon the plane normal to the flow direction and depending by the grid geometry.

K_B is the correlation factor accounting for the effect of grid porosity and degree of establishment of longitudinal flow inside the unit, depending on the ratio A_B/A_S and from the shell length to diameter ratio.

Experimental measures permitted to validate the general pressure drops correlations for different grid typologies. While correlations resulted in good agreement with measured data for high Re numbers and low viscosity fluids (i.e., for typical gas services) and for mid Re numbers (**Figure** 7), prediction at low Re numbers for viscous fluids, resulted less accurate, possibly suggesting the lower attitude of the grid as turbulence promoter while enveloped by a heavy viscous stream.

Indeed test-bench conducted in real Refinery environment in late 2004 permitted to assess lower fouling accumulation of same crude oil stream upon EMbaffle® unit, with respect to segmental unit, whereas both were placed in parallel on the same line.

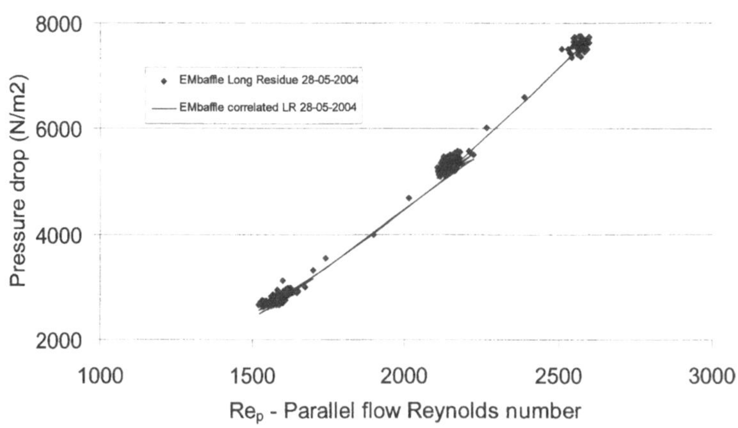

Figure 7.
Pressure drops as a function of Reynolds number for mid Re.

Further research performed by independent third party witnessed a potential for fouling reduction of five to seven times, in terms of deposited thickness over the same time frame. The consequences in terms of reduced maintenance and improved reliability descend straightforwardly.

The above anticipates some conclusions among applicable technologies. Longitudinal flow designs configure themselves as consistent alternative, whereas limited pressure drops, vibration issues, and fouling accumulation may deteriorate the performances of the unit; in this sense, their application directly descends from process requirements.

On the other side, as it is emerging with growing awareness, customization is a strong drive in the adoption of these technologies. In the case of EMbaffle®, the ability to govern fluid-dynamics by selecting the grid geometry and the number of baffles, permits to work on turbulence generated across the single baffles and the heat transferred as a consequence of it. Accordingly, EMbaffle® may be basically fitted-for-purpose, to an extent not achievable with segmental flow technologies, in which compromise between the thermal performance and the corresponding pressure drop has always to be searched.

Customization can be implemented for (i) reducing temperature approach, i.e., the active surface of the unit, then the dimensions and weight of the same, in some cases even the number of shells required for service; (ii) reducing bundle-related pressure drops, i.e., reducing the diameter of the unit and the thickness of the shell (especially valuable in case of high pressure services); (iii) optimizing the ratio of the heat transferred per units of pressure drops, by allowing optimized unit shaping for siting and power saving of routing pumps, compressors, and fans.

Tables 1 and **2** sum up the benefits categories of user experience in the adoption of the technology. **Table 1** concisely reports the key spillovers produced by the selected performance criteria; **Table 2** reports the impact exerted on the investments (CAPex) and O&M costs (OPex) and the relevance of the latter for the main stakeholders involved in the deal.

End users and EPCs represent the usual key decision players, hence ability to design-to-efficiency, -effectiveness, and other measures of performance allows the full valorization of the technology.

Measures of performance	Technological spillovers
Efficiency (temperature approach)	1. Compact dimensions 2. Reduced weight 3. Critical temperature services (i.e., close to freezing/ vaporization points or tight-range operational points)
Effectiveness (heat transferred per units of pressure drops)	4. Optimized shape 5. Reduced size and power consumption of pumps/compressors
Operating window (pressure drops)	6. Same unit for increased mass flow rates 7. Thinner shell/s under reduced size 8. Optimized maintenance of parallel lines
Reliability	9. Minimized vibrations 10. Reduced fouling deposition and accumulation 11. Reduced erosion of tubes bundle

Table 1.
Measures of performance.

	CAPex	OPex	End user investment dept	End user field and maintenance	EPC	Fabricator
1.	✓		✓		✓	✓
2.	✓		✓	✓	✓	✓
3.		✓		✓		
4.		✓		✓		
5.		✓		✓		
6.	✓	✓	✓	✓	✓	
7.	✓		✓		✓	
8.	✓		✓		✓	✓
9.		✓		✓		
10.	✓	✓	✓	✓		
11.	✓	✓	✓	✓		
12.	✓	✓	✓	✓		

Table 2.
Investment and operational costs perspective.

In this regard, comparison with reference conventional technology is suggested to the process engineer in order to promptly evaluate the convenience of alternate approach, based on the specific constraints raised from the field. Availability of EMbaffle® technology on HTRI design platform facilitates the preliminary job and may provide some useful insights over customization range for the specific service.

4. Applications and design cases

4.1 Gas dehydration

Wet gas, as extracted from subsea basin, may contain a relevant amount of water; part of it results in being bonded to the hydrocarbon blend and cannot be effectively removed through physical separation only. At pressures and temperatures well above the expected, water solidifies trapping the hydrocarbon inside a cage-like structure, known as hydrate. Hydrates may obstruct the section of the transportation pipeline and cause severe damage to the line and the equipment. One way to face this issue is to inject an effective inhibitor like mono-ethylene glycol (MEG) whose action is to absorb the water vapor.

Heat exchangers for gas dehydration service feature a MEG inlet manifold equipped with MEG sprayers at their ends for uniform distribution of the inhibiting agent inside the tubes where wet gas is routed and water content progressively removed. **Figure 8** illustrates typical spray particle distribution in the channel of the heat exchanger just upstream the tubes inlet. Heat for process completion is supplied by gas already dried, hence a feed-effluent design is in play.

Placed close to the gas extraction section, extremely high inlet gas pressures are common and content of vapors dictate adoption of suitable materials, being martensitic steels frequently used.

Conventional design for feed-effluent service and gas dehydration specifically is NTIW. Motivations rely on proved robustness and reliability; on the other side,

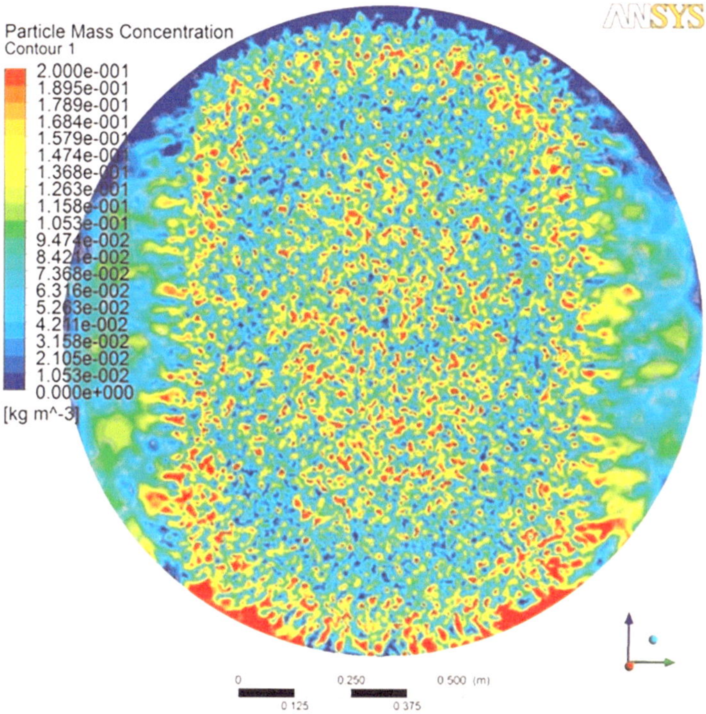

Figure 8.
Particle mass distribution of inhibiting agent. (Illustration rights by B&R.)

extended dimensions and inherent large pressure drops makes it less viable, whereas offshore platforms a/o floating vessels host the gas treatment facility and space and weight become critical factors.

EMbaffle® proves generally lighter than NTIW and better exploiting the allowable space while satisfying the process requirements (**Table 3**).

Item	Conventional design	Alternate design	Units
TEMA type	NEN	NEN	—
Number of equipment	3 parallel × 1 series	3 parallel × 1 series	—
Shell ID	1500	1350	mm
Tube length	13,000	13,000	mm
Baffle arrangement	NTIW	EMbaffle®	—
Installed area	14,589	13,740	m^2
SS pressure drops	1.0	1.0	bar
Duty	17,700	17,700	kW
Duty/installed area	1.213	1.288	kW/m^2
Weight	338.4	228.4	tons

Table 3.
Gas dehydration.

Furthermore, it proves valuable in reducing the unit length and mainly the shell diameter (in consideration of high pressures and thickness of the same).

4.2 Synthesis gas loop in ammonia production

Process heat exchangers (PHE) are used in synthesis gas loop in ammonia production, The Haber-Bosch process achieves conversion of hydrogen to ammonia in a catalyst-based converter; reaction rate is around 30%, hence multi-pass conversion suiting the loop-style approach is required.

Synthesis gas coming out from the Ammonia converter and rich in Ammonia is cooled to favor the subsequent separation of liquid Ammonia from unreacted hydrogen gas which is re-routed to the converter.

Large amount of heat carried by the synthesis gas after conversion is recovered by PHE.

No doubt that hydrogen conversion temperatures and pressures (typically 400–45°C at 140–220 bar) solicit attention to activation of corrosion mechanism, being hydrogen embrittlement, nitriding and hydrogen stress cracking the most recurrent ones.

Conventional, i.e., horizontal, layout of PHE is based on natural convection unit with separated steam drum on top of it and risers and downcomers connecting the two of them.

While keeping natural convection mechanism, i.e., natural flow of water to steam driven by buoyancy and its superior reliability, vertical U-tubes solutions had been developed aiming to a more compact design. In this regard, EMbaffle® offers an extremely simplified design with integrated inner steam drum at top of the shell.

U-tubes are arranged according to fountain lay-out; extensive use of low alloy grades in tube construction and ferrules in austenitic material for tube inlet prevent corrosion inducement.

Shell side boiling water, heated up by the tube-side synthesis gas, rises to the steam drum where liquid droplets are separated from vapor, which is generally routed, eventually following superheating step, to a utility line (for direct utilization or power production).

Due to open grid structure, EMbaffle® fits perfectly allowing full unconstrained cross and parallel free flow all along the vertical tube arrangement, promoting the homogeneously distributed density of boiling water/steam at any shell section level, with suppressed turbulence. The resulting extremely low pressure drops promote the highest recirculation factors, driving to increased conversion rates with reduced installed surface. On a performance basis, construction is the most simplified and cheap solution available in the market.

Innovation, in course of patenting, results quite simple as compared with generally complex alternate proprietary lay-outs and favors lean construction and assembly.

Figure 9 illustrates a 5 years' operating unit in North America ammonia plant.

Advantages in terms of reduced footprint, weight, and complexity emerge: savings to CAPex (due to compact size and lighter weight) and OPex (impact of weight upon transportation and siting, while reduced complexity means higher reliability) do follow.

4.3 Nitric acid production

Nitric acid is a main precursor in the production of inorganic fertilizers of commercial cut; Ostwald process achieves the oxidation of ammonia in a catalytic reactor.

Figure 9.
EMbaffle® arrangement for process gas boiler.

Reactor releases nitrogen monoxide which is cooled in a heat exchanger line (HEL) and further oxidized to nitrogen dioxide before being routed to the absorption column where is converted in nitric acid by addicting water. Unconverted nitrogen dioxide and impurities (a blend referred as tail gas) are routed back to the HEL to cool the nitrous gas.

Materials of construction are challenging in consideration of corrosive attack of nitrous gas, especially at tube inlet, where it comes below its dew point: austenitic 18/10 grades may be therefore prescribed for tube and even shell construction.

HEL is a quasi-feed effluent exchanger line, made of multiple units in series, known as tail gas preheaters: Rod Baffle layout is a consolidated design, as common in any feed effluent requests (under limited pressure drops and no request for additional input of heat). Request for improved competitiveness move the process engineer to identify technological solutions improving overall effectiveness by keeping the advantages of longitudinal flow layout.

EMbaffle® proves valuable in reducing the unit length and, mostly, the shell diameter (in consideration of high pressures required and thinner thickness of the same), under same process prescriptions. Reduction in weight and footprint further facilitate transportation of the unit/s to site and siting in existing facilities.

Advantages in terms of reduced footprint, weight, and complexity emerge: savings to CAPex (due to thinner thickness, more compact size, and lighter weight) and OPex (reduced complexity means higher reliability and low maintenance demand) do follow.

Table 4 provides comparison against conventional (Rod Baffle) layout under same process requirements.

4.4 Naphtha hydrotreating

Heavy naphtha must be (hydro)treated before being reformed in the catalytic unit to remove sulfur, hydrogen a/o metals. Lighter fractions are treated too, before being

Iitem	Conventional design	Alternate design	Units
TEMA type	NEN	NEN	—
Number of equipment	1 parallel × 1 series	1 parallel × 1 series	—
Shell ID	2080	1745	mm
Tube length	8700	7030	mm
Baffle arrangement	Rod Baffle	EMbaffle®	—
Installed area	2544	1819	m^2
Overall pressure drops	9.0	9.0	kPa
Duty	8373	8373	kW
Duty/installed area	3.291	4.603	kW/m^2
Weight	60.9	39.4	tons

Table 4.
Nitric acid production.

cracked and reduced to shorter chains for producing distillate blends (e.g., diesel fuel, kerosene or even gasoline further to catalytic reforming).

Role of hydro-treating in refinery is therefore hard to overestimate.

Hydro-desulphurization (HDS) is a major class of hydro-treating processes, aiming at reducing the sulfur content in the naphtha stream before it is routed to the catalytic reformer: scope is to preserve the catalyst from poisoning, produce commercial naphtha cut and reduce the environmental impact of the same.

Separation of sulfidic acid from the hydrocarbon chain occurs in the converter at 280–420°C at high partial pressures of hydrogen; heat recovery units are required to recover the heat in the sulfur-free naphtha effluent and supply it to the fresh charge, in a typical feed effluent layout.

Materials of construction are critical due to hydrogen sulfide combined with hydrogen at high temperature and partial pressure: low alloy steels based on molybdenum and chromium are widely used to face thermal creep and hydrogen embrittlement; yet, whereas design temperatures are lower carbon steel may be used.

Segmental S&T exchangers are used in consideration of their robustness and reliability; yet, in case of modernization projects addressing increase in processed mass flow rates on both sides, whereas limitations exist for accommodation of novel units, E to F TEMA layout shift may provide a valuable hint to designer. Shift shell is checked against additional pressure drops which may exceed the limitations of the segmental unit, as it does occur frequently.

Longitudinal flow designs provide one additional drive to adoption of F layout. Equivalence of pressure levels across the longitudinal baffle, due to very limited pressure drops along the exchangers, favors its adoption; benefits range from reduced unit length to easier accommodation of novel units in existing areas.

Table 5 provides comparison of EMbaffle® against conventional (single segmental) solution under same process requirements; need for extended exchanger line length penalizes the pressure drops, the number of units and the overall footprint.

Item	Conventional design	Alternate design	Units
TEMA type	AES	AFS	—
Number of equipment	2 parallel × 8 series	2 parallel × 4 series	—
Shell ID	660	1067	mm
Tube length	9144	9144	mm
Baffle arrangement	Single segmental	EMbaffle®	—
Installed area	4450	3791	m^2
SS pressure drops	9.32	1.90	bar
Duty	68,000	68,000	kW
Duty/installed area	15.28	17.94	kW/m^2
Weight	—	—	tons

Table 5.
Naphtha HDS.

4.5 Thermal storage in renewable power plants

In the first decade of the twenty-first century, concentrated solar power plants paved the way toward a new approach in utility scale power production. The concept is that solar heat radiation can be gathered and used for producing high temperature water steam for power production. A suitable intermediate fluid (HTF) carries the solar heat and releases it to water for producing steam. The drawback is that sun is not available throughout the entire day and during cloudy periods; more-over consumption loads may require some flexibility in power delivery.

In order to decouple and store the energy generated by the plant, in the so called Parabolic Through layout, process engineers conceived a smart system made of two large tanks filled with molten salts, selected as preferred thermal storage material. During the sunny day hot HTF, exceeding the production request, heats up the molten salt stored in the cold tank and routed to the hot tank; during late day hours and in cloudy days hot molten salt heats up the HTF before being routed to the cold tank.

A set of heat exchangers transfer heat between the molten salts and the HTF in a daily cycling process. Early technology was based on available segmental units; double segmental was preferred due to lower pressure drops under the typical large mass flow rates of molten salts, generally routed on shell side, also to simplify recovery in case of salt freezing during the night.

The solution proved robust but expensive: for standard European 50 MW plant size, six large exchangers were required to accomplish the duty.

Replacement of double segmental with EMbaffle® permitted to halve this figure (**Figure 10**), by leveraging on its superior effectiveness, i.e., same heat transferred over lower pressure drops. Technology, proved in several plant configurations in Europe, North and South Africa, demonstrated its adaptability to different requirements in terms of transportation to site, siting constraints and logistic requests.

Advantages in terms of reduced number of units and complexity emerge: savings to CAPex (due to reduced weight) and OPex (due to impact of weight upon transportation and siting, while reduced complexity means higher reliability) do follow.

Figure 10.
EMbaffle® units in thermal storage service in a CSP facility.

Item	Conventional design	Alternate design	Units
TEMA type	BFU	BFT	—
Number of equipment	6 series	3 series	—
Shell ID	2860	2130	mm
Tube length	10,400	21,000	mm
Baffle arrangement	Double segmental	EMbaffle	—
Installed area	27,646	22,025	m^2
SS pressure drops	3.0	2.1	bar
Duty	134,580	134,580	kW
Duty/installed area	4.87	6.11	kW/m^2
Weight	810	595	tons

Table 6.
Thermal storage in CSP plants.

Table 6 provides comparison of EMbaffle® against better conventional (double segmental) solution under same process requirements.

5. Conclusions

Growth in demand of energy solicits development of new technologies, compliant with tighter environmental friendly requirements and motivated by awareness of progressive scarcity of fossil fuel resources. S&T heat exchangers, among the draft horses in the industry, have always been part of this process. Longitudinal flow technology bloomed in the second half of the last century to address specific issues, later becoming a comprehensive solution in view of its customization-ability, whereas conventional (segmental) design failed or simply resulted out of the balance. EMbaffle® technology emerged as a valuable option, addressing diverse process engineering and field requests in green-field projects and even brown-field replacements of existing units. Technology may provide lower pressure drops and higher heat transferred per unit length of the exchanger, larger mass flow rates for same pressure drops, tighter temperature approaches, reduced fouling accumulation, and no-inception of vibration on the bundle. Yet, these benefits translate in terms of more compact sizing, reduced footprint, and weight with positive impacts on CAPex and OPex. It's not surprising that several services in the gas, petrochemical, and chemical industries took profit from it. Feed effluent services did stand out as prominent application field: in the gas treatment and purification sector, EMbaffle® proves limited pressure drops drive the choice of the technology, whereas in the catalytic reforming sector, the driver is the high partial pressure of the hydrogen stream which recommends the containment of the unit diameter. Finally, in the chemical and fertilizing industry, large mass flow rates of feed charges make full bore technology largely preferred over alternatives. Moving from historical referrals in crude oil preheating and Over'd condensation, EMbaffle® waded in the gas sector supplying large inter-cooling units for on-shore compression pipelines and later off-shore compression facilities aboard FLNG vessels. Confidence achieved in these developments triggered innovative concepts, actually embedded in standard designs. Working closely in touch with Process Licensors and Engineering Co., novel solutions are identified, and improvements are elaborated on solid bases. Upon these premises, EMbaffle® technology will continue to pave a remarkable way in the industry.

Nomenclature

A	heat transfer area (m^2)
A_B	Baffle flow area (m^2)
A_{iM}	EMbaffle grid projected area (m^2)
A_R	ring area (m^2)
A_s	shell flow area (m^2)
c_p	specific heat capacity (J/kg K)
C_L	laminar heat transfer geometry function (−)
C_T	turbulent heat transfer geometry function (−)
CSP	concentrated solar power (−)
D_h	characteristic diameter for Nu and Re_h (m)
D_P	characteristic diameter for Re_P (m)
D_S	shell inner diameter (m)
D_T	tube outer diameter (m)
f_F	fanning friction factor (−)
EPC	engineering procurement & construction
h	shell-side film transfer coefficient (W/m^2K)
HTF	heat transfer fluid
K_b	hydraulic loss coefficient of baffle (−)
L_T	length of tubes (m)
N_B	number of baffles (−)
N_T	number of tubes (−)
Nu	Nusselt number ($= \frac{h}{k} D_h$) s(−)

Pr	Prandtl number $(= \frac{c_p \mu_b}{k})$ (–)
P_T	tube pitch (m)
Re_h	heat transfer Reynolds number (–)
Re_p	parallel flow Reynolds number (–)
U_B	baffle velocity calculated from A_b (m/s)
U_s	shell-side velocity calculated form A_s (m/s)
ΔP	pressure drop (Pa)
ΔP_B	baffle flow pressure drop (Pa)
ΔP_L	longitudinal flow pressure drop (Pa)
k	shell-side fluid thermal conductivity (W/m K)
μ_b	bulk viscosity (Pa s)
μ_w	wall viscosity (Pa s)
ρ	mass density (kg/m^3)

Author details

Marcello Garavaglia*, Fabio Grisoni, Marta Mantegazza and Marco Rottoli
Brembana&Rolle SpA, Valbrembo, Italy

*Address all correspondence to: mgaravaglia@brembanarolle.com

References

[1] Thulukkanam K. Heat Exchanger Design Handbook. 2nd ed. Boca Raton: CRC Press; 2013

[2] Heat Exchangers Types. Available from: https://heat-exchanger-world.com/tema-standards-and-the-shell-and-tube-heat-exchanger-design [Accessed: August 28, 2023]

[3] Tubular Exchanger Manufacturers Association, Inc. Standards of the Tubular Exchanger Manufacturers Association. 10th ed. 2019

[4] Twisted Tubes® Heat Exchanger. Available from: https://www.kochheattransfer.com/products/twisted-tube-bundle-technology [Accessed: August 28, 2023]

[5] Helixchanger® Technology. Available from: https://www.lummustechnology.com/catalysts-and-equipment/heat-transfer-equipment/helixchanger-heat-exchanger [Accessed: August 28, 2023]

[6] Gentry CC. Rod baffle heat exchanger design methods. In: The VI Intl. Symposium on Transport Phenomena in Thermal Engineering South Korea. 1993

[7] Small WM. Heat Exchanger Selection - Part III - RODbaffle Heat Exchangers. HTFS; 1987

[8] Rottoli M, Odry T, Agazzi D, Notarbartolo E. EMbaffle® heat exchanger technology. In: Innovative Heat Exchangers. Switzerland: Springer Book; 2016. pp. 341-362

[9] Rottoli M, Agazzi D, Garavaglia M, Grisoni F. EMbaffle® heat transfer technology - step up in CO_2 reduction. In: Heat Transfer - Design, Experimentation and Applications. London: IntechOpen; 2021

[10] Perrone F, Brignone M, Micali G, Rottoli M. Grid geometry effects on pressure drops and heat transfer in an EMbaffle® heat exchanger. In: International CAE Conference 2014, Verona, Italy. 2014

[11] Oshinowo L, Kuhn D. Turbulence decay behind expanded metal screens. The Canadian Journal of Chemical Engineering. 2000;**78**:1032-1039

[12] Ansari A. Influence of expanded metal screen on downstream turbulent flow [MSc thesis]. Department of Chemical Engineering. Canada: University of Toronto; 1998

[13] Smits AJ. Viscous Flow and Turbulence – Lectures in Fluid Mechanics. United Kingdom: University of Princeton; 2009

[14] Ishiyama EM et al. Management of crude preheat trains subject to fouling. Heat Transfer Engineering. 2013;**34**: 692-701

[15] HTRI. Available from: https://www.htri.net/

Chapter 5

Heat Transfer Mechanisms in Petroleum and Geothermal Wellbores

Sharat V. Chandrasekhar, Udaya B. Sathuvalli and Poodipeddi V. Suryanarayana

Abstract

The flow of fluids between wells and reservoirs involves a substantial amount of thermal energy exchange with the formation. Understanding the mechanisms involved in the heat transfer of these processes is crucial to the design of the wells for mechanical integrity. While long term production scenarios may achieve a notional steady state, short term injection scenarios involve an accurate consideration of the thermal transients. With global initiatives towards a transition to clean energy, the design of geothermal wells is becoming an area of great importance these days. Accordingly, correct simulation of the heat transfer in the circulating scenario involved in closed loop wells enables accurate assessments of thermal power generated. This chapter aims to educate the user in how to tackle these problems and explains the physics and mathematics involved in detail.

Keywords: heat transfer, wellbores, production, injection, circulation, geothermal energy

1. Introduction

1.1 Background

The discovery of reservoirs with hotter in-situ temperatures (above 200° F) over the past several decades has introduced engineering challenges that depend critically on an accurate assessment of wellbore temperatures. In particular, subsea wells are being drilled to deeper horizons these days and are exposed to hotter temperatures than in the past.

These wells have multiple tubulars and fluid-filled annuli as depicted in **Figure 1**. In addition, many of these wells are prolific producers (of hydrocarbons or geothermally heated water), resulting in high arrival temperatures at the surface. In some instances, the fluid arrival temperatures at the wellhead could, in fact, be hotter than the already high bottomhole temperature, because of the negative Joule-Thomson effect. A problem of equal, if not greater importance, is the effect of the lateral (or radial) heat transfer from the flowing stream to the wellbore layers, resulting in

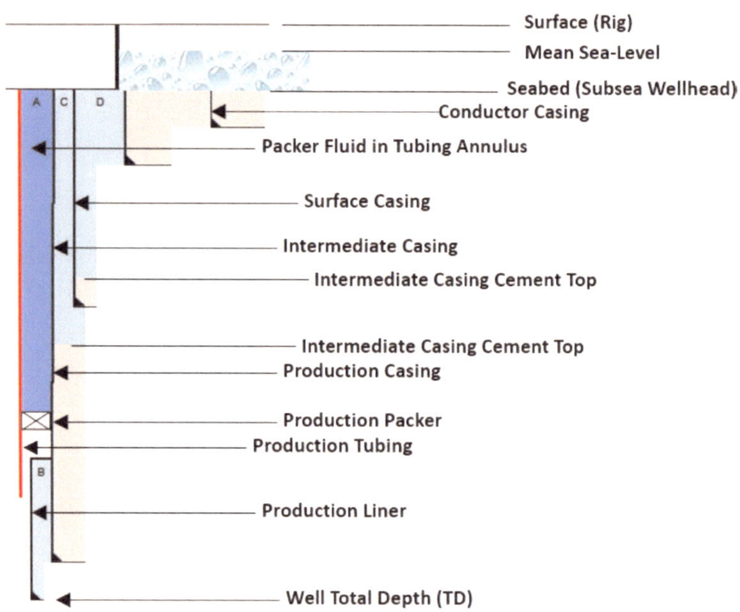

Figure 1.
Schematic of a complex wellbore with multiple annuli and bounding tubulars.

temperature buildup in fluid filled annuli and thermal stresses in the unsupported sections of the bounding tubulars.[1] One of the most serious implications of radial heat transfer is Annular Pressure Buildup (APB). The prediction and mitigation of APB constitutes a vast body of investigation in its own right. Thermal stresses in tubulars influence the structural design of the wellhead, and the control of Wellhead Movement (WHM). The displacement constraints on the tubulars at the wellhead and the tops of cement can cause buckling and the generation of bending stresses during well operation. In a worst case discharge (WCD) scenario, elevated temperatures may potentially dislodge tubulars from the wellhead, and require additional lock down rings to prevent the tubulars from catapulting. All of these phenomena require accurate and reliable estimates of wellbore temperatures. In instances that involve operations with short durations, accurate prediction of the thermal transient response is critical (for example. Drillstem tests, Well Testing to evaluate reservoir performance, Designing APB mitigation mechanisms, wellhead pressure control in platform wells). Injection and circulation scenarios also create temperature changes that generate unsustainable tensile forces in improperly designed wellbore tubulars and tubular connections.

1.2 Heat transfer mechanisms in wellbores

The fundamental mechanisms of heat transfer in a wellbore are indicated in **Figure 2**. In most of the onshore and offshore locations, the geothermal temperature increases with depth below the surface, at an average of rate of 21–32°C/km. This

[1] Wellbore casings (see **Figure 1**) are hollow cylinders with diameter to wall thickness ratios between 8 and 40. These hollow cylinders are known as Oil Country Tubular Goods (OCTG) or tubulars.

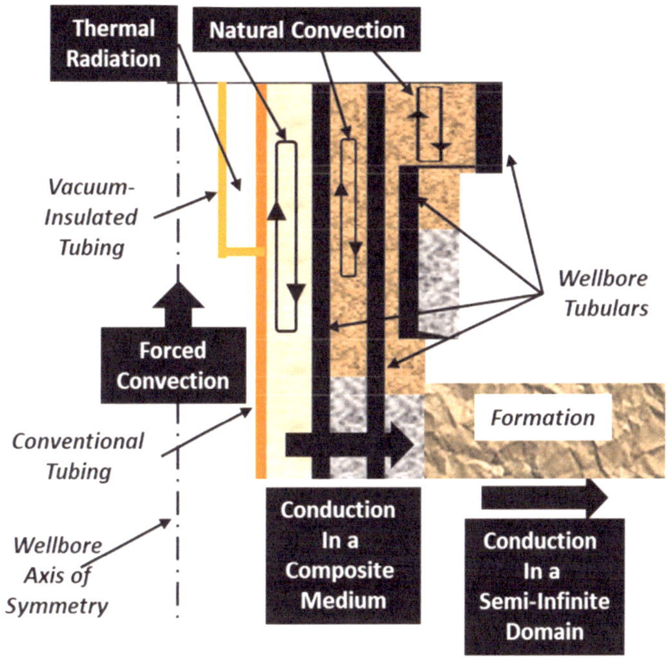

Figure 2.
Illustration of the various heat transfer phenomena in a wellbore.

temperature gradient is the primary driver for all heat exchange phenomena in a wellbore. This is true of wellbores used to extract oil and gas, and of wellbores used to generate geothermal energy.

For the purposes of thermal and structural analysis, a well can be enclosed in an imaginary volume that encloses the production tubing (i.e. the innermost cylinder and primary flow conduit), and the series of casings and cement sheaths in the intervening annular spaces. The well boundary is located at the interface between the outermost cement sheath and the earth (known hereafter as the formation).

In a wellbore, energy is exchanged between the flow stream(s), the wellbore (i.e. the casing strings and annular contents) and the formation. The thermal analysis of the producing wellbore proceeds in three interlinked steps. The first step is the solution of the balance (mass, momentum, and energy) equations in the tubing. The second step is the assessment of radial heat loss from the tubing to the wellbore. For the purposes of thermal analysis, the wellbore is defined as the region between the outer surface of the tubing and the outer surface of the outermost cement sheath. The third step is the determination of the heat transfer in the formation, i.e. from the wellbore – formation boundary to the earth.

There is forced convection heat transfer between the flowing fluid stream and the conduit boundary. Usually, the uncemented annular sections between tubulars contain incompressible fluids that experience natural convection. Conduction in the radial direction occurs through the walls of the casing, and the cemented sections of the intervening annuli. This is a case of diffusion across in a composite medium. At the well boundary, heat lost by the contents of the wellbore diffuses by conduction into a semi-infinite domain. Sometimes the semi-infinite domain is approximated by a finite domain with a very large farfield radius. In some wells, there is a need to minimise heat loss from the wellbore. In such applications, Vacuum Insulated tubing (VIT) is used.[2] The heat transfer in this case between the inner and outer pipes is practically by thermal radiation.

1.3 Types of well thermal operations

In terms of thermal interactions, a wellbore is essentially a heat exchanger. Conventional heat exchangers typically involve heat transfer between two counterflowing or parallel streams. In a wellbore however, a single stream flowing up (production) or down (injection) the wellbore, exchanges heat with the formation layers surrounding the wellbore, as indicated in the left two panels of **Figure 3**. In this figure, the black dotted line represents the geothermal temperature, which prevails in the wellbore until an operation (or operations) induce a thermal disturbance. During production, the hot fluid exits from the reservoir at the bottom of the well and flows upward. During it upward transit, there is loss of fluid enthalpy because of lateral/radial heat transfer. This is responsible for the heating of the tubulars, the annular contents in the well (solid red curve, panel (a)). During injection, cold fluid gets heated during its downward transit (blue curve, panel (b)). The right two panels indicate circulation scenarios which are analogous to classic counterflow heat exchangers. In both cases, qualitative descriptions of the associated temperature profiles are indicated (solid red and blue curves).

In all the three scenarios in **Figure 3**, the key objective of a thermal analysis is the prediction of the flowing temperature profiles, given appropriate boundary conditions. In the case of transient heat transfer, initial conditions must also be specified. In production and injection scenarios, the (boundary condition) temperatures are either known or stipulated at the reservoir and wellhead locations. In forward circulation, the temperature is specified at the wellhead location of the inner conduit, whereas in reverse circulation the temperature is specified at the wellhead location of the outer (annular) conduit. In either case, the temperature is specified at the inlet to the wellbore of the downward flowing stream. At the bottom of the wellbore, it is typical (but not always) to stipulate the equality of the temperatures of the two flowing streams, as shown in **Figure 3** (panels (**c**) and (**d**)).

The analysis involves the solution of the transport equations in conjunction with heat transfer in the formation. This requires careful consideration of all relevant fluid and thermal transport phenomena. The subsequent sections will present a systematic analytical approach to the solution of the aforementioned problems.

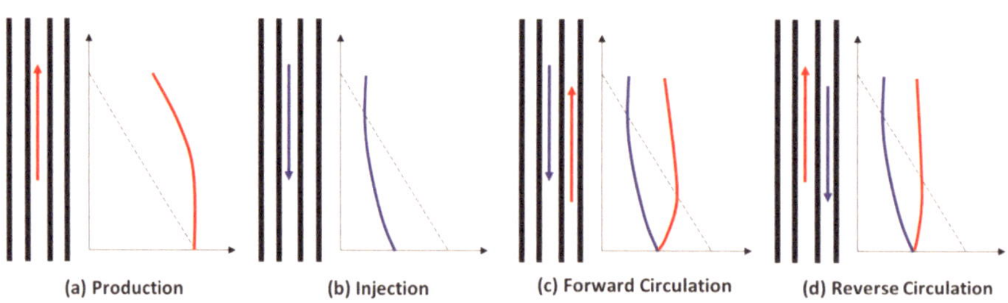

Figure 3.
Producing (a), injection (b), forward circulating (c), and reverse circulating (d) scenarios.

[2] A joint of VIT contains a set of concentric pipes welded together at the ends of the shorter tube. The annular gap between the pipes is evacuated to ~20 millitor (2.6 Pa).

1.4 Review of relevant literature

The earliest studies of heat transfer in wellbores by Lesem et al. [1] and Moss and White [2] date back to the late 1950s. For a detailed review of the literature on the topic, the reader is referred to Chandrasekhar [3] wherein a comprehensive transient thermal model of a complex wellbore is described in detail. There are several key studies that constitute essential reading and are listed below:

The 1962 study by Ramey [4] was the first systematic study of both flowing and wellbore temperatures. His approach assumed pseudo steady state conditions in the flowing conduit and wellbore, with the transients relegated solely to the formation. This approach is in fact the basis for a very large number of model implementations (the WELLFLO code for example) to this day. The wellbore itself was modelled as a line source in a semi-infinite formation for which a simple expression was used to characterise the transient heat flux. While the approach breaks down for shorter producing intervals, it is valid for time periods corresponding to Fourier numbers in excess of unity.

Willhite [5] extended the approach of Ramey [4] to account for amongst other phenomena, natural convection, and thermal radiation in fluid-filled annuli. An iterative approach is required to calculate the overall heat transfer coefficient linking the temperature of flowing stream to the far field undisturbed geothermal temperature.

It is very likely that Raymond [6] was the first study of the transient circulation problem using a combine Laplace Transform/Finite Difference approach. The key observation of Raymond's analysis is that the transients are limited to the first few hours of circulation and that the steady state solution was valid for longer periods. The first detailed study of multiple well operating scenarios is that of Wooley [7] in which production, injection, and circulation were considered in the context of a transient analysis using a finite difference approach to couple the well and formation responses.

More recent studies have looked at analytical solutions where possible for coupled wellbore/formation problems. Wu and Pruess [8] considered transient heat transfer between a flowing fluid stream and the formation, but used an overall lumped heat transfer coefficient to model the heat transfer across the wellbore itself. They formulated a more refined formation temperature model using Laplace transforms to model a cylindrical source. The 2018 study of Chandrasekhar et al. [9] is recommended for the reader interested in the application of a circulating model to a complex realistic wellbore considering both hydraulics and thermal phenomena, in addition to several other aspects of actual real-life wellbores.

There are several textbooks in the literature that present a detailed analysis of the fundamentals of wellbore heat transfer. Hasan and Kabir [10] cover several aspects of both heat transfer and fluid flow in wellbores, starting with the governing equations, and several models for multiphase flows in wellbores. In a 2009 SPE monograph, Mitchell and Sathuvalli [11] discuss various phenomena and analytical techniques relevant to temperature prediction in prolific oil and gas producers.

There are a few experimental studies that have investigated aspects of wellbore heat transfer. Jones [12] describes a real time measurement that was quite novel at the time approach to establish circulating temperatures in wellbores during drilling and cementing operations. The performance of Vacuum-Insulated Tubing was studied by Aeschliman et al. [13] in the context of a steam injection well. Their results compared six different commercially available means of achieving thermal insulation by the suppression of convection in the tubing annulus.

2. Governing transport equations

2.1 Mass conservation

Consider a control volume (CV) of length Δz and a fixed radius R as shown in **Figure 4**. Mass, momentum, and energy enter and leave the CV at locations z and $z + \Delta z$. In wellbores, usually there is no mass accumulation at a given location in the control volume, so that the constant mass flow rate is given by

$$\dot{m} = \rho(z)V(z) = \rho(z + \Delta z)V(z + \Delta z)A \qquad (1)$$

where $A = \pi R^2$ is the conduit flow area. A mass balance over the control volume in the limit that the size Δz shrinks to zero yields

$$\frac{\partial}{\partial z}(\rho V) = -\frac{\partial \rho}{\partial t} = 0 \qquad (2)$$

so that the instantaneous temporal derivative of the density is zero. From Eqs. (1) and (2), we have

$$\frac{\partial V}{\partial t} = \frac{\partial}{\partial t}\left(\frac{\dot{m}}{\rho A}\right) = \frac{\dot{m}}{A}\frac{\partial}{\partial t}\left(\frac{1}{\rho}\right) = -\frac{V}{\rho}\frac{\partial \rho}{\partial t} = 0 \qquad (3)$$

whereupon the temporal derivative of the velocity also vanishes, so that along with no mass accumulation, there is no accumulation of momentum in the control volume either.

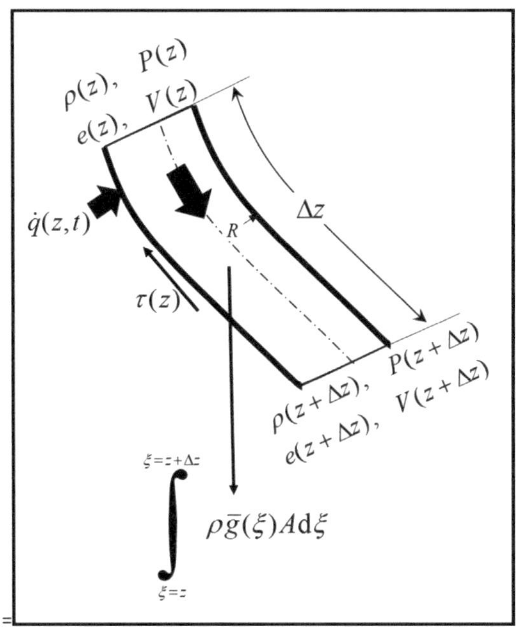

Figure 4.
Control volume for mass, momentum, and energy balances.

2.2 The energy equation

The specific energy $e(z,t)$ identified in the figure is the sum of the kinetic, potential, and internal energies such that

$$e = u + \frac{1}{2}V^2 - \delta g y \tag{4}$$

where y is the vertical depth relative to some fixed datum, and the negative sign associated with it implies a loss of potential energy as the vertical depth increases. The term $\delta = \pm 1$ defines the orientation of the streamwise coordinate relative to the gravity vector, such that $\delta = 1$ and $\delta = -1$ describe injection and production scenarios, respectively. The issue of how to deal with these terms in a circulation scenario will be described later.

Energy enters and leaves the CV in **Figure 4** as indicated, with some accumulation at the rate Δe over a period Δt. Energy is supplied through the conduit boundary at the rate $\dot{q}(t)$.

A simple energy balance yields the following expression

$$\dot{m}[e(z+\Delta z) - e(z)] + \rho A \Delta z \frac{\Delta e}{\Delta t} = 2\pi R \int_{s=z}^{s=z+\Delta z} \dot{q}(s)\mathrm{d}s - \dot{m}\Delta\left(\frac{P}{\rho}\right) = (2\pi R \Delta z)\dot{q}(z + \lambda \Delta z, t) \tag{5}$$

where the Mean Value Theorem has been used to replace the integral such that $0 < \lambda < 1$. Dividing by Δz and taking the limit as both $\Delta z \to 0$ and $\Delta t \to 0$ results in

$$\lim_{\substack{\Delta z \to 0 \\ \Delta t \to 0}} \left\{ \dot{m}\left[\frac{e(z+\Delta z) - e(z)}{\Delta z}\right] + \rho A \frac{\Delta e}{\Delta t} \right\} = \lim_{\substack{\Delta z \to 0 \\ \Delta t \to 0}} \{2\pi R \dot{q}(z + \lambda \Delta z)\} \tag{6}$$

Evaluating the limit and noting that $\dot{m} = \rho A V$ simplifies Eq. (6) to

$$\rho \frac{\partial e}{\partial t} + \rho V \frac{\partial e}{\partial z} = \frac{2}{R}\dot{q}(z) \tag{7}$$

From one of the fundamental thermodynamic relationships relating the enthalpy to internal energy, we have

$$h = u + \frac{P}{\rho} \tag{8}$$

Substitution of the above along with Eq. (4) into Eq. (7) yields

$$\rho\left[\frac{\partial u}{\partial t} + V\cancel{\frac{\partial V}{\partial t}}^{0} + \cancel{\frac{\partial}{\partial t}}(\delta g y)^{0}\right] + \rho V\left(\frac{\partial h}{\partial z} + V\frac{\partial V}{\partial z} - \delta g\right) = \frac{2}{R}\dot{q}(z,t) \tag{9}$$

The derivative of the vertical depth y with the streamwise coordinate z (known as the *Measured Depth* in wellbore parlance) is the cosine of the local wellbore inclination θ. From Eq. (8), the temporal derivative of the internal energy can be expressed as

$$\frac{\partial u}{\partial t} = \frac{\partial h}{\partial t} - \frac{\partial P}{\partial t}\frac{}{\rho} = \frac{\partial h}{\partial t} - \left[\frac{P}{\rho}\frac{\partial \not{P}^0}{\not{\partial} t} - \frac{P}{\rho^2}\frac{\partial \not{\rho}^0}{\not{\partial} t}\frac{P}{\rho}\right] = \frac{\partial h}{\partial t} \tag{10}$$

where the temporal derivative of the density vanishes in accordance with the constant mass flow criterion which also stipulates from Eq. (3) that the time derivative of the velocity is zero. In addition, it follows from the momentum conservation equation (which will be presented in the next section), that in the context of a constant mass flow rate the time derivative of the pressure also vanishes. Accordingly, Eq. (9) reduces to

$$\rho\frac{\partial h}{\partial t} + \rho V\left(\frac{\partial h}{\partial z} + V\frac{\partial V}{\partial z} - \delta g\cos\theta\right) = \frac{2}{R}\dot{q}(z,t) \tag{11}$$

Note that Eq. (11) contains spatial derivatives of both specific enthalpy and velocity. Closure therefore requires the consideration of the momentum equation which is presented next. It is reiterated here that Eq. (11) as derived is ***only*** valid under the assumption of a constant mass flow rate throughout the wellbore.

2.3 The momentum equation

A force balance over the same control volume as in **Figure 4** yields the rate of change of momentum. The forces acting on the fluid in the control volume are the static and dynamic pressure forces and the pressure and shear stress as indicated

$$\begin{aligned} &A\left[\rho(z+\Delta z)V^2(z+\Delta z) - \rho(z)V^2(z)\right] + A[P(z+\Delta z) - P(z)] + \rho A\Delta z\frac{\not{\Delta V}^0}{\Delta t} \\ &= \delta\rho\overline{g}(z)A\Delta z + 2\pi R\Delta z\int_{s=z}^{s=z+\Delta z}\tau(s)\mathrm{d}s = \delta\rho\overline{g}(z)A\Delta z + (2\pi R\Delta z)\tau(z+\lambda\Delta z) \end{aligned} \tag{12}$$

whereupon following the same logic as was used to derive the energy equation and noting from Eq. (3) that the time derivative of the velocity vanishes, we have

$$\rho V\frac{\partial V}{\partial z} + \frac{\partial P}{\partial z} = \delta\rho g\cos\theta + \frac{2}{R}\tau(z) \tag{13}$$

which is functionally equivalent to Newton's Second Law of Motion relating the rate of change of Momentum to the sum of the forces acting on a body of fluid.

Note that the stipulation of no mass accumulation also implies no momentum accumulation, so that the only accumulation in the wellbore is that of energy. Note also, that from Eq. (13), the time derivative of pressure is zero which enables the energy equation to be cast with enthalpy as the sole flux variable on both side of the equation.

2.4 Coupled transport equation system

The kinetic energy term in Eq. (11) is represented by the spatial derivative of the velocity. This term can be expressed in terms of pressure and enthalpy derivatives by invoking the chain rule as follows

$$\begin{aligned}\frac{\partial V}{\partial z} &= \frac{\partial}{\partial z}\left(\frac{\dot{m}}{\rho A}\right) = \frac{\dot{m}}{A}\frac{\partial}{\partial z}\left(\frac{1}{\rho}\right) = -\frac{V}{\rho}\frac{\partial \rho}{\partial z} \\ &= -V\left[\left(\frac{1}{\rho}\frac{\partial \rho}{\partial P}\bigg|_h\right)\frac{\partial P}{\partial z} + \left(\frac{1}{\rho}\frac{\partial \rho}{\partial h}\bigg|_P\right)\frac{\partial h}{\partial z}\right] = V\left[\alpha_h\frac{\partial h}{\partial z} - \beta\frac{\partial P}{\partial z}\right]\end{aligned} \tag{14}$$

where $\beta = \frac{1}{\rho}\frac{\partial \rho}{\partial P}\big|_h$ is the adiabatic compressibility and $\alpha_h = -\frac{1}{\rho}\frac{\partial \rho}{\partial h}\big|_P$ can be regarded as a two-phase isobaric volume expansivity.[3] Substitution of Eq. (14) into Eqs. (11) and (13) yields the system of equations that can be expressed in the compact form

$$\begin{aligned}&\rho\frac{\partial}{\partial t}\begin{bmatrix} h \\ 0 \end{bmatrix} + \begin{bmatrix} \rho V(1+V^2\alpha_h) & -\rho V^3\beta \\ \rho V^2\alpha_h & 1-\rho V^2\beta \end{bmatrix}\frac{\partial}{\partial z}\begin{bmatrix} h \\ P \end{bmatrix} \\ &= \begin{bmatrix} 2R^{-1}\dot{q}(z,t) + \delta\rho V g\cos\theta \\ \delta\rho g\cos\theta + 2R^{-1}\tau(z) \end{bmatrix}\end{aligned} \tag{15}$$

2.5 Extraction of wellbore temperatures

Subject to an initial condition for the enthalpy field in the wellbore and appropriate pressure and enthalpy and boundary conditions, Eq. (15) can be solved in conjunction with the constitutive models for the heat flux ($\dot{q}(z,t)$) and frictional resistance $\tau(z)$ terms. Once the enthalpy and pressure distributions are known, the temperature distribution is determined from the appropriate thermophysical property database[4] or a correlation that has the functional form

$$T = T(h, P) \tag{16}$$

Since the heat flux term itself depends on temperature, the solution involves an iterative sequence at each depth. Furthermore, in a transient multiphase analysis, the coupling of the transport equations with the diffusion in the formation adjacent to the wellbore can present occasional challenges with respect to the latter, as is the case with modelling transient phenomena in steam injector wells.

2.6 Single phase flow (sensible heat)

In Two-Phase flow, the temperature remains constant under an isobaric change in enthalpy. In single phase flow however, an enthalpy change is related to changes in pressure and temperature according to

$$\mathrm{d}h = c_\mathrm{p}\mathrm{d}T - c_\mathrm{p}c_\mathrm{JT}\mathrm{d}P \tag{17}$$

where c_p and c_JT are the specific heat at constant pressure and the fluid Joule-Thomson Coefficient, respectively. The latter is related to the fluid volume expansivity according to

[3] Note that the volume expansivity is typically defined as the normalised density derivative with respect to temperature.

[4] Such as NIST's REFPROP.

$$c_{\mathrm{JT}} = \frac{1}{\rho c_{\mathrm{p}}}(\alpha T - 1) \tag{18}$$

For liquids with very low expansivity, the Joule-Thomson coefficient is invariably negative. The term "sensible heat" is used to refer to Eq. (17) since a change in enthalpy can be perceived as a change in temperature, which is not possible when the state point is inside the vapour dome of the fluid.

The sensible heat formulation can also be extended to multiphase flow in wellbores in conjunction with the so-called *Black-Oil Model*, where weighted properties are used for the fluid thermophysical properties in each of the phases. The key advantage of Eq. (17) is that the primary flux variables are now pressure and temperature. Accordingly, the spatial velocity derivative is now expressed as

$$\frac{\partial V}{\partial z} = -V\left[\left(\frac{1}{\rho}\frac{\partial \rho}{\partial P}\bigg|_T\right)\frac{\partial P}{\partial z} + \left(\frac{1}{\rho}\frac{\partial \rho}{\partial T}\bigg|_P\right)\frac{\partial T}{\partial z}\right] = V\left[\alpha\frac{\partial T}{\partial z} - \beta\frac{\partial P}{\partial z}\right] \tag{19}$$

where $\alpha = -\frac{1}{\rho}\frac{\partial \rho}{\partial T}\big|_P$ is the (single-phase) isobaric volume expansivity, and $\beta = \frac{1}{\rho}\frac{\partial \rho}{\partial P}\big|_T$ is the isothermal (not adiabatic) compressibility.

Substitution of Eq. (19) in Eq. (15) results in the system (see Chandrasekhar [3]):

$$\rho c_{\mathrm{p}}\frac{\partial}{\partial t}\begin{bmatrix} T \\ 0 \end{bmatrix} + \begin{bmatrix} \rho V\left(c_{\mathrm{p}} + V^2\alpha\right) & -\rho V\left(c_{\mathrm{p}}c_{\mathrm{JT}} + V^2\beta\right) \\ \rho V^2\alpha & 1 - \rho V^2\beta \end{bmatrix}\frac{\partial}{\partial z}\begin{bmatrix} T \\ P \end{bmatrix} = \begin{bmatrix} Q + VH \\ H + F \end{bmatrix} \tag{20}$$

where

$$\begin{aligned} Q_{\mathrm{tbg}} &= 2R^{-1}\dot{q}(z,t) = 2\pi\frac{\overline{U}}{\dot{m}_{\mathrm{tbg}}}\left(T_{\mathrm{ann}} - T_{\mathrm{tbg}}\right) \\ H_{\mathrm{tbg}} &= -g\cos\theta\rho_{\mathrm{tbg}}V_{\mathrm{tbg}} \\ F_{\mathrm{tbg}} &= 2R^{-1}\tau(z) = \frac{1}{2D}f\rho V^2 \end{aligned} \tag{21}$$

are the Thermal, Hydrostatic, and Frictional forcing functions respectively. If we ignore the transient term for the time being, then the 2×2 system in Eq. (20) can be inverted to yield the following expressions for the streamwise gradients of temperature and pressure as

$$\frac{\mathrm{d}T}{\mathrm{d}z} = \frac{(1-\omega)Q + (\alpha TV)H + (\alpha T + \omega - 1)VF}{\rho V\left[(1-\omega)\left(c_{\mathrm{p}} + \eta\alpha\right) + \eta(\alpha T + \omega - 1)\alpha\right]} \tag{22}$$

and

$$\frac{\mathrm{d}P}{\mathrm{d}z} = \frac{c_{\mathrm{p}}H + \left(c_{\mathrm{p}} + \eta\alpha\right)F - \left(g_{\mathrm{c}}{}^{-1}V\alpha\right)Q}{\left[(1-\omega)\left(c_{\mathrm{p}} + \eta\alpha\right) + \eta(\alpha T + \omega - 1)\alpha\right]} \tag{23}$$

where

$$\eta = V^2 \tag{24}$$

and

$$\omega = \rho V^2 \beta = \eta \rho \beta \tag{25}$$

For an incompressible liquid, $\alpha = \beta = \omega = 0$, and accordingly the expression for the temperature gradient reduces to

$$\frac{\mathrm{d}T}{\mathrm{d}z} = \frac{Q - VF}{\rho V c_\mathrm{p}} \tag{26}$$

and the pressure gradient reduces as it should, to

$$\frac{\mathrm{dP}}{\mathrm{d}z} = H + F \tag{27}$$

Before delving into the important physical aspects of Eq. (26), it will be useful to establish the constitutive models for the heat flux and fluid shear stress, which will be presented next.

3. Constitutive models

3.1 Heat flux

The formation adjacent to a wellbore is notionally a semi-infinite cylinder. Accordingly, a true steady state is never reached. However at large times from when a well is put into operation, a notional or *pseudo-steady state* condition is reached as shown by Ramey [4]. Under these conditions, the heat flux between the flowing fluid stream in the wellbore and the formation can be represented in terms of the local temperature difference between the fluid and the undisturbed formation temperature prevailing at a distance far from the wellbore, at any given depth. Mathematically this can be represented by the very simple form

$$\dot{q}(z) - -U\left[T(z) - T_{\mathrm{gco}}(z)\right] \tag{28}$$

where U which will be described in more detail to follow, is an *Overall Heat Transfer coefficient* that is independent of time, and is associated with the conduit radius R, as reflected by the $2\pi R$ term in Eq. (5). Note that the factor of 2π itself has already been incorporated in Eq. (5) and therefore does not appear in Eq. (28).

3.2 Overall heat transfer coefficient

Consider the section of the wellbore shown in **Figure 5** with multiple intervening layers between the fluid and the formation. At some notional steady state typically attained at long elapsed times after a well is put into operation, the fluxes across all of these layers are equal. In addition, this flux is also equal to the flux at the wellbore-formation interface at some frozen time instant t. This assumption corresponds to what is termed a *Pseudo-Steady-State* approach. With respect to the nomenclature of the figure, the heat flux per unit length (not unit area) is

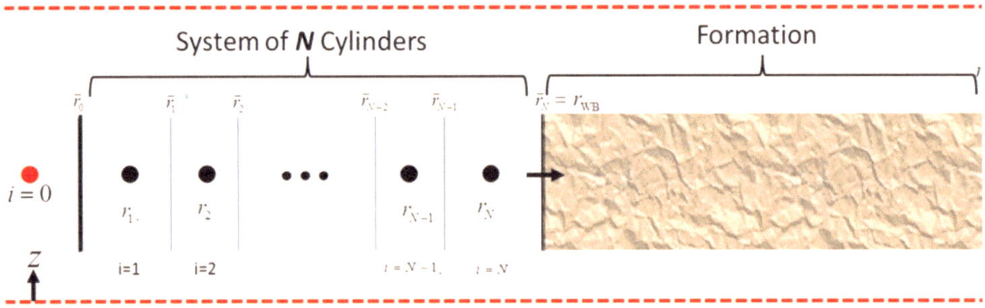

Figure 5.
Wellbore layers between the transport fluid and the formation.

$$\frac{\dot{q}_L(z)}{2\pi} = -hR\left(T_0 - \overline{T}_0\right) = -\frac{k_1}{\ln \bar{r}_1/\bar{r}_0}\left(\overline{T}_0 - \overline{T}_1\right) = -\frac{k_2}{\ln \bar{r}_2/\bar{r}_1}\left(\overline{T}_1 - \overline{T}_2\right) = \cdots$$
$$= -\frac{k_j}{\ln \bar{r}_j/\bar{r}_{j-1}}\left(\overline{T}_{j-1} - \overline{T}_j\right) = \cdots = -\frac{k_N}{\ln \bar{r}_N/\bar{r}_{N-1}}\left(\overline{T}_{N-1} - \overline{T}_N\right) \tag{29}$$
$$= -R_{\text{wb}}k_{\text{Geo}}\frac{\partial T}{\partial r}(t)\bigg|_{r=R_{\text{wb}}} = -k_{\text{Geo}}\left(\overline{T}_N - T_{\text{Geo}}(z)\right)\frac{\partial \theta}{\partial \eta}(\tau)\bigg|_{\eta=1}$$

where the dimensionless time $\tau = \frac{\alpha_{\text{Geo}}}{R_{\text{wb}}^2}t$ is a Fourier Number. In Eq. (29) the barred entities refer to interface locations (layer boundaries) and the subscript 0 in the first term on the RHS of Eq. (29) refers to the fluid. Note that this term describes forced convection between the fluid and the conduit. The dimensionless flux in the last term of Eq. (29) is independent of the wellbore outer radius, interface temperature, and formation properties and is obtained from the solution of the diffusion problem in a cylindrical semi-infinite domain. Ramey [4] presented an expression for the dimensionless flux in terms of the Fourier Number[5] based on an approximation of the line source solution as

$$\frac{\partial \theta}{\partial \eta}(\tau)\bigg|_{\eta=1} = F(\tau) = -\left(\ln \frac{1}{2\sqrt{\tau}} + 0.29\right)^{-1} \tag{30}$$

The constant heat flux per unit length across the wellbore represented by Eq. (29) and out into the formation at some snapshot in time can also be represented in terms of a fluid to formation temperature difference with the use of an Overall Heat Transfer Coefficient as

$$\frac{\dot{q}_L(z)}{2\pi} = UR(T - T_{\text{Geo}}) \tag{31}$$

Eliminating the flux $\dot{q}(z)$ and the temperatures between Eqs. (29) and (31) results in the following expression for the overall heat transfer coefficient

[5] Note that Ramey's solution is not accurate for small values of the Fourier Number. An expression for $F(\tau)$ that is valid over the entire spectrum of Fourier Numbers is provided in [3].

$$U = \frac{1}{R}\left[\frac{1}{hR} + \sum_{k=1}^{k=N}\frac{\ln {}^{r_j}\!/_{r_{j-1}}}{k_j} + \frac{1}{k_{\mathrm{Geo}}F(\tau)}\right]^{-1} \tag{32}$$

Eq. (32) considers the following phenomena.

- forced convection in the conduit,
- thermal resistances offered by all of the intervening wellbore layers,
- thermal resistance at the wellbore-formation interface.

It is often convenient in wellbore heat transfer analysis to work with an overall *conductance* rather than a coefficient. In the context of Eq. (32), this is defined as $\overline{U} = UR$, so that Eq. (28) can be rewritten as

$$\dot{q}(z)R = -\overline{U}\left[T_0(z) - T_{\mathrm{geo}}(z)\right] \tag{33}$$

where the overall conductance is the reciprocal of the term in brackets in Eq. (32).

3.3 Natural convection in fluid-filled annuli

The thermal conductivity in each wellbore layer depends on the medium of the layer. In the case of tubulars and cemented sections, the thermal conductivity may be regarded as constant. Typical values for steel and cement are 45 W/m-K and 1 W/m-K, respectively. When the layer consists of a fluid however, it is subject to natural convection that must be considered in the analysis. Therefore, the conductivity of a fluid layer as used in Eq. (32) should be replaced by an equivalent thermal conductivity k_{eq} that accounts for natural convection. In terms of a heat transfer coefficient the flux due to natural convection between the inner and outer walls of the layer is given by

$$\dot{q}_i(z) = h_i\bar{r}_{i-1}\left(\overline{T}_{i-1} - \overline{T}_i\right) = \frac{k_{\mathrm{eq}}}{\ln {}^{\bar{r}_i}\!/_{\bar{r}_{i-1}}}\left(\overline{T}_{i-1} - \overline{T}_i\right) \tag{34}$$

The natural convection correlation used in this context is an extension of the one proposed by Dropkin and Sommerscales [14] as suggested by Willhite [5] such that the equivalent conductivity k_{eq} of the layer can be obtained by using a multiplier on the thermal conductivity of the static medium that corresponds to the Nusselt Number from the Dropkin-Somerscales correlation according to

$$\frac{k_{\mathrm{eq}}}{k_i} = \frac{h_i\bar{r}_{i-1}}{k_i} = \mathrm{Nu} = 0.049(\mathrm{GrPr})^{{}^1\!/_3\mathrm{Pr}^{0.074}} \tag{35}$$

where the Grashof and Prandtl numbers are defined as

$$\mathrm{Gr} = \frac{\beta_{\mathrm{f}}g\left(\overline{T}_{i-1} - \overline{T}_i\right)\left(\bar{r}_{i-1} - \bar{r}_i\right)^3}{\nu^2} \tag{36}$$

and

$$\mathrm{Pr} = \frac{\nu}{\alpha} \tag{37}$$

where in the interest of keeping with the traditional nomenclature used in the literature, the term β_f in Eq. (36) is the coefficient of volumetric thermal expansion (essentially the isobaric volume expansivity), and should not be mistaken for the isothermal compressibility.[6] Owing to the exponent of one-third in Eq. (35), an iterative procedure is required to evaluate the overall heat transfer coefficient, with Eqs. (32)–(37) all embedded in the iterative loop.

3.4 Shear stress

For flow in a conduit, the frictional resistance is expressed as a shear stress per unit distance in the streamwise gradient that is related to the flow velocity according to the Darcy–Weisbach model

$$\tau(z) = -\frac{1}{8} f \rho V^2 \tag{38}$$

where the friction factor $f = f(\mathrm{Re}, {}^{\varepsilon}\!/_{D})$ can be obtained in terms of the flow Reynolds number and the pipe roughness (ε) to diameter ratio, according to the iterative Colebrook-White model or any one of several noniterative approximations published in the literature. Note that the negative sign in Eq. (38) implies that the shear stress acts in the direction opposing the flow.

4. Steady state temperature profiles

Consider the scenario depicted in **Figure 6**, in which fluid enters a vertical wellbore from a reservoir at a fixed temperature T_{BH}. This temperature is generally referred to as the *Static Bottomhole Temperature*. The formation temperature is assumed to decrease linearly with depth down to T_{Surf} at the wellbore exit, such that

$$T_{geo}(z) = T_{BH} - \frac{z}{L}(T_{BH} - T_{Surf}) \tag{39}$$

Dimensionless streamwise coordinate, and fluid and geothermal temperatures can be defined according to

$$\xi = \frac{z}{L}, \quad \theta = \frac{T - T_{Surf}}{T_{BH} - T_{Surf}}, \quad \theta_{geo} = \frac{T_{geo} - T_{Surf}}{T_{BH} - T_{Surf}} = 1 - \xi \tag{40}$$

Substitution of Eqs. (28) and (38) for the heat flux and shear stress into Eq. (26) and rearranging, results in the compact form

$$\frac{d\theta}{d\xi} = N_{TU}(\theta_{geo} - \theta) + \Lambda = N_{TU}(1 - \xi - \theta) + \Lambda \tag{41}$$

[6] This rather unfortunate reusing of symbols in context is somewhat typical of heat transfer analysis, when a multitude of thermal phenomena are considered. Note that α can refer to both volume expansivity and thermal diffusivity.

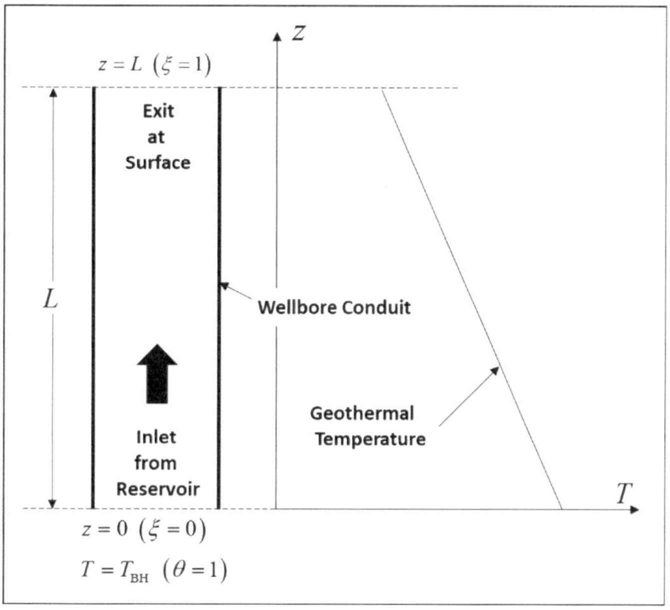

Figure 6.
Production through a wellbore from a reservoir.

where readers familiar with classical heat exchanger analysis will identify the coefficient of the temperature differential as the *Number of Transfer Units* defined in this context as

$$N_{\mathrm{TU}} = 2\frac{URL}{\rho V R^2 c_{\mathrm{p}}} = 2\pi\frac{\overline{U}L}{\dot{m}c_{\mathrm{p}}} \equiv \frac{1}{\mathrm{Pe}} \tag{42}$$

which as noted above can be expressed as the reciprocal of a[7] Peclet Number. The dimensionless *Frictional Heating* parameter in Eq. (41) is defined as

$$\Lambda = \frac{f}{4}\left(\frac{V^2}{c_{\mathrm{p}}(T_{\mathrm{BH}} - T_{\mathrm{Surf}})}\right)\frac{L}{R} \tag{43}$$

Note that the dimensionless entity in parenthesis within the expression for Λ is the *Eckert Number*. If the temperature at the inlet to the wellbore is the same as the reservoir temperature T_{BH}, the boundary condition corresponding to Eq. (41) is

$$\theta(0) = 1 \tag{44}$$

Subject to the boundary condition above, the solution of Eq. (41) is

$$\theta(\xi) = 1 - \xi + Ce^{-N_{\mathrm{TU}}\xi} + \frac{1+\Lambda}{N_{\mathrm{TU}}} \tag{45}$$

For $N_{\mathrm{TU}} = 1$, the dimensionless temperature profiles are plotted in **Figure 7** for various values of the frictional heating parameter Λ. What is noteworthy is that as Λ

[7] The use of the article a in "a Peclet Number" as opposed to the Peclet Number is because there are several flavours of this entity relating the magnitudes of the advective to thermal diffusive fluxes.

increases from zero, the temperature at the surface (known as the *Arrival Temperature*) not only approaches the reservoir temperature, but in fact exceeds it, a very common observation in prolific deepwater oil producers. Neglecting the frictional heating term can therefore result in a severe underprediction of temperatures and threaten wellbore integrity if the attendant tubular thermal stresses and annular pressure buildup are accordingly underpredicted.

The impact of the Number of Transfer Units is shown in **Figure 8**, and shows that even in the absence of frictional heating, near-isothermal conditions in the wellbore

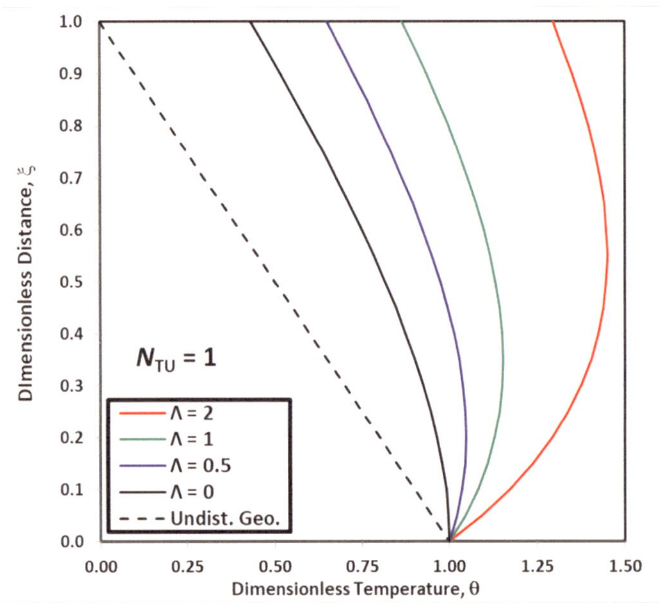

Figure 7.
Temperature profiles for N_{TU} = 1 *and various values of* Λ.

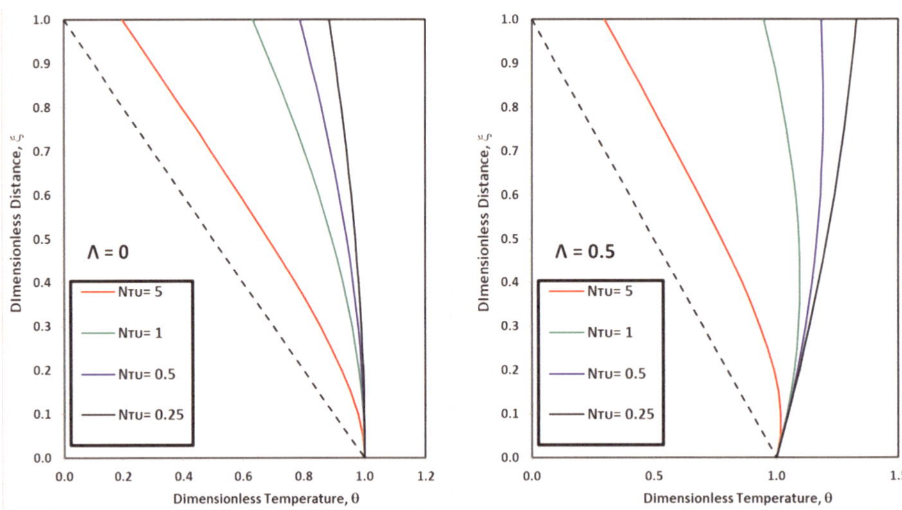

Figure 8.
Effect of N_{TU} *on temperature profiles for* Λ = 0 *(left) and* Λ = 0.5 *(right).*

can be achieved as the Number of Transfer Units number becomes very small. In fact, in the limit $N_{TU} \rightarrow 0$, the RHS of Eq. (41) reduces to zero for $\Lambda = 0$ implying $\theta(\xi) = 1$ throughout in accordance with the boundary condition of Eq. (44).

A word of caution is in order here. The phenomenon of the arrival temperature exceeding the bottomhole temperature as evidenced in **Figure 7** is associated only with liquids that almost always have a Negative Joule-Thomson coefficient c_{JT}. For gases, c_{JT} is negative only below the inversion pressure. Accordingly, for very high gas rate flows in wellbores, it is not uncommon to see the contrary effect of a substantial drop in the fluid temperature towards the surface. Attempting to simulate this effect however, with negative values of Λ (positive c_{JT}) will yield results which while seemingly plausible may not be accurate since Eq. (41) is only valid for incompressible flows, and the assumptions invoked in its derivation tend to break down when the produced fluid is predominantly gaseous.

5. Transient heat transfer in wellbores

Thermal transients in a wellbore are characterised by the fluid exchanging heat with the surrounding formation at a rate that evolves in time. Therefore, there are two adjacent coupled problems that need to be considered – the transient transport equation in the wellbore conduit of radius R and the transient radial diffusion in the formation. These two problems are coupled at the interface between the outer layer of the wellbore and the formation at the radius $\overline{R}$ as shown in **Figure 9**. Between the radial locations R and $\overline{R}$ are all of the wellbore layers comprised of tubulars and annuli. For the purpose of this illustrative example, it will be assumed that these layers have

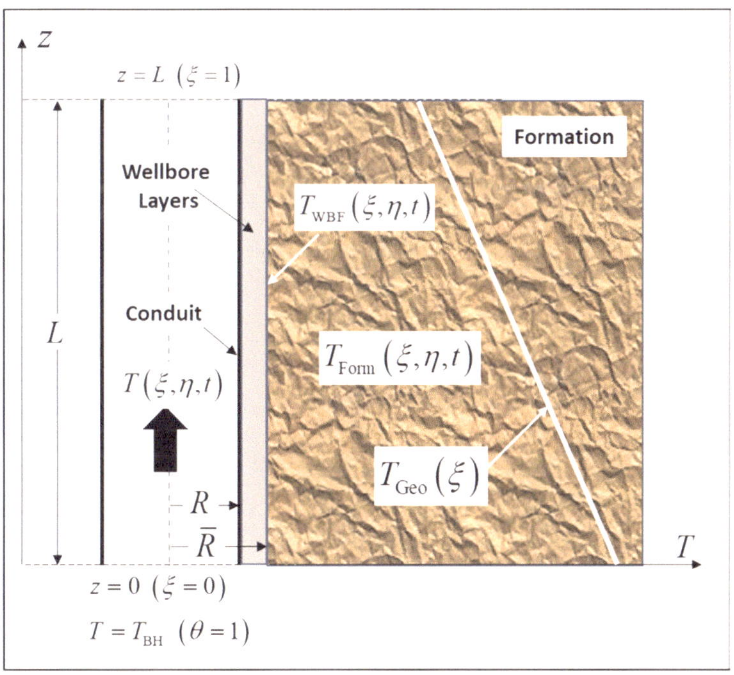

Figure 9.
Transient production through a wellbore.

negligible capacitance, so that they respond instantaneously to the fluid transients.[8] Unlike in the steady state case where the overall heat transfer coefficient across the wellbore layers was used to link the fluid temperature to the undisturbed geothermal temperature, in a transient analysis, the linkage is between the fluid temperature and the transient temperature at the wellbore-formation interface according to

$$\dot{q}(z,t) = -U[T(z,t) - T_{\text{WBF}}(z,t)] = k_{\text{Geo}} \left.\frac{\partial T_{\text{Form}}}{\partial r}\right|_{r=\overline{R}} \tag{46}$$

With the assumption of an incompressible fluid, the transient transport equation can be extracted from Eq. (20) as

$$\frac{\partial T}{\partial t} + V\frac{\partial T}{\partial z} = \frac{Q - VF}{\rho c_{\text{p}}} \tag{47}$$

subject to the same boundary condition as in the steady state case i.e., $T(0,t) = T_{\text{BH}}$ and the initial condition

$$T(z,0) = T_{\text{geo}}(z) \tag{48}$$

The temperature at the interface between the wellbore and the formation is not known a-priori, but constitutes one of the radial boundary conditions for the problem of diffusion in the formation. In lieu of a semi-infinite domain, we will regard the formation as a finite cylindrical domain with an outer radius far enough that geothermal conditions prevail therein at the end of the well operational time period of interest. Accordingly, the diffusion in the formation is governed by the partial differential equation

$$\rho_{\text{geo}} c_{\text{p}_{\text{geo}}} \frac{\partial T_{\text{Form}}}{\partial t} = k_{\text{geo}} \frac{\partial^2 T_{\text{Form}}}{\partial r^2} \tag{49}$$

subject to the initial condition

$$T_{\text{Form}}(r,0,z) = T_{\text{Geo}}(z) \tag{50}$$

and the boundary conditions

$$T_{\text{Form}}\left(\overline{R},\tau,z\right) = T_{\text{WBF}}(z) \tag{51}$$

and

$$\frac{\partial T_{\text{Form}}}{\partial r}\left(\overline{R}_{\infty},\tau,z\right) = 0 \tag{52}$$

Implicit in Eq. (49) is the assumption that axial diffusion is negligible, which given the length scales of typical wellbores, is eminently justified. As a consequence, the governing equation holds at all depths along the wellbore where the thermal interaction between the wellbore and the formation is described in terms of a purely radial heat transfer mechanism.

---

8 For an analysis that considers the thermal transients in all layers of a complex wellbore, see [3].

The time and spatial coordinates variables are non-dimensionalised as

$$\xi = \frac{z}{L}, \quad \eta = \frac{r}{\overline{R}}, \quad \eta_\infty = \frac{R_\infty}{\overline{R}}, \quad \tau = \frac{\alpha_{\text{geo}}}{\overline{R}^2} t \tag{53}$$

where the dimensionless time is essentially a Fourier Number, and the temperatures are normalised as before with respect to the bottomhole to surface geothermal temperature difference as

$$\begin{aligned} \theta &= \frac{T - T_{\text{Surf}}}{T_{\text{BH}} - T_{\text{Surf}}} & \theta_{\text{form}} &= \frac{T_{\text{Form}} - T_{\text{Surf}}}{T_{\text{BH}} - T_{\text{Surf}}} \\ \psi &= \frac{T_{\text{WBF}} - T_{\text{Surf}}}{T_{\text{BH}} - T_{\text{Surf}}} & \phi &= \frac{T_{\text{Form}} - T_{\text{Surf}}}{T_{\text{BH}} - T_{\text{Surf}}} \end{aligned} \tag{54}$$

Substitution of the modified heat flux constitutive model i.e., Eq. (46), the shear stress model from Eq. (38) and the set of dimensionless variables defined by Eqs. (53) and (54) into Eq. (47) and Eqs. (48)–(52) result in the following dimensionless system of coupled equations:

$$\frac{\partial \theta}{\partial \tau} + \text{Pe}\frac{\partial \theta}{\partial \xi} = \Gamma(\psi - \theta) + \Lambda \tag{55}$$

where

$$\text{Pe} = \frac{V\overline{R}}{\alpha_{\text{geo}}}\frac{\overline{R}}{L} \tag{56}$$

is a Peclet number. In the context of the steady state problem, the term

$$\Gamma = 2\frac{\overline{R}}{R}\frac{U\overline{R}}{\rho c_{\text{p}}\alpha_{\text{geo}}} \tag{57}$$

is a diffusion coefficient and the frictional heating parameter

$$\Lambda = \frac{f}{4}\frac{\overline{R}}{R}\frac{V^3\overline{R}}{c_{\text{p}}\alpha_{\text{geo}}(T_{\text{BH}} - T_{\text{Surf}})} \tag{58}$$

as defined above is somewhat different from that described earlier. Eq. (55) is subject to the initial condition

$$\theta(\xi, 0) = 1 - \xi \tag{59}$$

and the boundary condition

$$\theta(0, \tau) = 1 \tag{60}$$

which must be solved in conjunction with the formation diffusion problem non-dimensionalised as

$$\frac{\partial \phi}{\partial \tau} = \frac{\partial^2 \phi}{\partial \eta^2} \tag{61}$$

subject to the initial condition

$$\phi(\eta, \xi, 0) = 1 - \xi \tag{62}$$

along with the a priori unknown boundary condition at the wellbore formation interface

$$\phi(1, \xi, \tau) = \psi(\xi, \tau) \tag{63}$$

and the farfield boundary condition

$$\frac{\partial \phi}{\partial \eta}(\eta_\infty, \xi, \tau) = 0 \tag{64}$$

The value of the farfield radius ratio η_∞ must be choses so as to be consistent with the physics of the problem. While the Neumann condition (zero flux) in Eq. (64) at the drainage radius is by itself adequate from a mathematical standpoint to provide closure to the system of equations, physical realism also requires that the formation temperature asymptotically approach the undisturbed geothermal temperature at the depth in question, at a radial location prior to the drainage radius. Failure to satisfy this criterion due to an insufficiently large value of η_∞ could result in substantially inaccurate calculations. A good rule of thumb for estimating the required drainage ratio is $\eta_\infty = 5\sqrt{\tau}$ where the Fourier number corresponds to the end of the time period of interest.

5.1 Solution of the transient formation diffusion problem

We start with the solution of Eq. (61) subject to the criteria of Eqs. (62)–(64) that following Ozisik [15], involves the use of Duhamel's Theorem as is customary for problems involving time dependent boundary conditions. The radial temperature profile is not the actual entity of interest. What is necessary to facilitate the coupling of the formation diffusion problem with the fluid transport equation, is the interface flux wherein flux continuity as expressed by Eq. (46) yields in terms of non-dimensional entities

$$\psi - \theta = \gamma \frac{\mathrm{d}\phi}{\mathrm{d}\eta}(\xi, \tau)\bigg|_{\eta=1} = \gamma \sum_{j=1}^{\infty} \overline{C}_j D_j(\xi, \tau) \tag{65}$$

where

$$\gamma = \frac{k_{\mathrm{Geo}}}{U\overline{R}} \tag{66}$$

and the Duhamel Convolution Integral is

$$D_j(\xi, \tau) = -\int_0^\tau e^{-\lambda_j^2(\tau-\beta)} \psi'(\xi, \beta) \mathrm{d}\beta \tag{67}$$

where the prime denotes a derivative with respect to β. The eigenvalues λ_j and the Fourier-Bessel coefficients $\overline{C}_j$ are defined in the appendix and depend on the farfield radius ratio η_∞. In practice, the infinite summation in Eq. (65) is obviously restricted to a finite number of Fourier modes. Note that the purpose of the exercise above was to express the temperature difference $\psi - \theta$ in terms of an interface flux.

5.2 Solution of the transient fluid transport equation

As is the case with a large number of problems in transient heat transfer, the first step in the solution of Eq. (55) is Laplace transformation whereupon we have

$$s\Theta - \theta(\xi, 0) + \mathrm{Pe}\frac{d\Theta}{d\xi} = \Gamma(\Psi - \Theta) + \frac{\Lambda}{s} \tag{68}$$

Laplace transformation of the interface flux expression of Eq. (65) in conjunction with Eq. (67) and some algebra yields

$$\Psi - \Theta = -\gamma \sum_{j=1}^{\infty} \overline{C}_j \left[\frac{s\Psi - \psi(0, \xi)}{s + \lambda_j^2}\right] \tag{69}$$

where the Convolution Theorem has been used on the Duhamel Integral. Noting that the initial condition or the interface flux is the undisturbed geothermal temperature, Eq. (69) can be rearranged (see [3]) into the compact form

$$\Psi - \Theta = G(s)(1 - \xi) - sG(s)\Theta \tag{70}$$

where

$$G(s) = \frac{W(s)}{1 + sW(s)} \tag{71}$$

and

$$W(s) = \gamma \sum_{j=1}^{\infty} \frac{\overline{C}_j}{s + \lambda_j^2} \tag{72}$$

Substitution of Eq. (70) into Eq. (68) along with the initial condition of Eq. (59) results in the ordinary differential equation in the frequency domain

$$\frac{d\Theta}{d\xi} = \left(\frac{1 + \Gamma G}{\mathrm{Pe}}\right)(1 - \xi) - s\left(\frac{1 + \Gamma G}{\mathrm{Pe}}\right)\Theta + \left(\frac{\Lambda}{\mathrm{Pe}}\right)s^{-1} \tag{73}$$

subject to the transformed boundary condition

$$\Theta(0, s) = \frac{1}{s} \tag{74}$$

The solution of Eq. (73) in the frequency domain is

$$\Theta(\xi, s) = F(s)\left(1 - e^{-A(s)\xi}\right) + \frac{1}{s}\left(e^{-A(s)\xi} - \xi\right) \tag{75}$$

where

$$F(s) = \frac{1}{s} + \left(\frac{\mathrm{Pe} + \Lambda}{1 + \Gamma G(s)}\right)\frac{1}{s^2} \tag{76}$$

and

$$A(s) = s\left(\frac{1 + \Gamma G(s)}{\mathrm{Pe}}\right) \tag{77}$$

5.3 Inversion from the frequency domain

The solution expressed in Eq. (75) in the frequency domain is not particularly useful from a practical point of view. It must therefore be inverted back to physical space using the Inverse Laplace Transform. One approach is to use the Cauchy Residue Theorem by summing the residues over all of the poles of Eq. (75). One of these poles is at the Origin.[9] The remaining poles lie along the negative real axis of the complex plane and are the zeros of the denominator of the 2nd term in Eq. (76) such that

$$f(s) = 1 + \Gamma G(s) = 0 \tag{78}$$

which must be solved numerically. An efficient method of doing so involves an asymptotic bracketing technique, the description of which is outside of the scope of this chapter.

The Residue Theorem while attractive from the standpoint of constituting a formal analytical solution can involve some very tedious if otherwise straightforward book-keeping in addition to the requirement of numerical evaluation of the roots of Eq. (78). A far more efficient approach is to use numerical inversion with the Gaver-Stehfest Function-Sampling Algorithm (Stehfest [16]) whereupon the temperatures in physical space are given by

$$\theta(\xi,\tau) \approx -\frac{\ln 2}{\tau}\sum_{k=1}^{2N_G}\sigma_k\Theta\left(\frac{k\ln 2}{\tau}\right) \tag{79}$$

where N_{GS} is the (even) order of the Gaver Summation, and $\sigma_k, k = 1\cdots N_{GS}$ are the Stehfest Accelerators, defined and listed in **Table 1** in the Appendix.

The evolution of the temperature profiles for two cases –with and without frictional heating, is shown in **Figure 10**. In both cases, the Gaver-Stehfest function sampling algorithm was used in conjunction with a farfield radius ratio of 200 and 1000 Fourier modes. The Negative Joule-Thomson effect is clearly seen in the right panel of the figure for the case with frictional heating.

[9] There is also a double-pole at the origin on account of the s^{-2} term in Eq. (76).

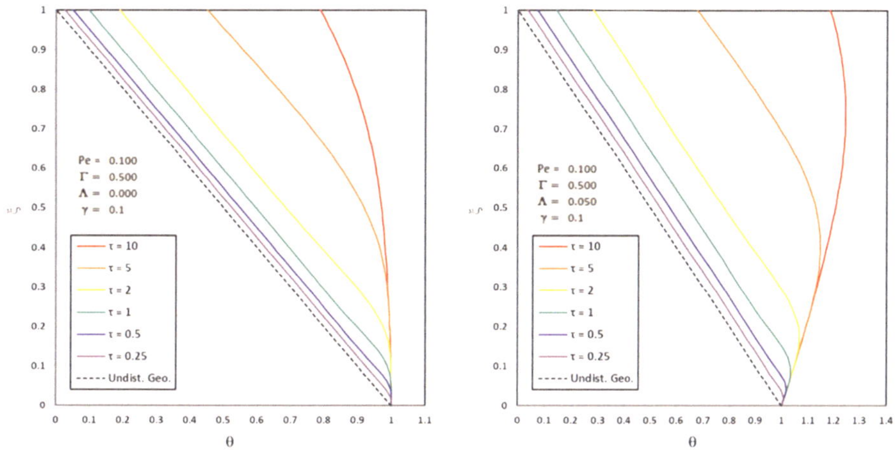

Figure 10.
Evolution of the transient temperature profiles for $\Lambda = 0$ *(left) and* $\Lambda = 0.05$ *(right).*

6. Heat transfer in circulating scenarios

Circulation constitutes an important aspect of wellbore operations. This is most commonly encountered in drilling, as well as swapping fluids and hole cleaning. In what is known as forward circulation, fluid is pumped down a drillstring and returns to the surface through the annulus, as depicted in panel (c) of **Figure 3**. In Reverse Circulation (panel d **Figure 3**), the flow directions are reversed so that colder fluid is injected down the annulus and hotter fluid returns to the surface. This is not that common in conventional wellbores, but is the primary mode of operation in geothermal wells where hot fluid returns through an insulated or partially insulated inner string known in that context as the tubing.

The thermal interactions in both scenarios are depicted in **Figure 11** indicating the known boundary conditions and the a priori unknown return fluid temperatures of interest. At the bottom of the well, a matching condition stipulates that the pipe and annulus temperatures (designated by the subscripts "p" and "a", respectively) are equal. In the simple well configuration considered, there is an inner pipe of inner radius R. The annulus has inner and outer radii R_i and R_o as indicated. The outer casing is cemented with the outer radius of the wellbore R_{WB} in contact with the formation. A linear geothermal gradient is assumed.

Invoking the previous assumptions of both incompressibility and pseudo steady state heat transfer, the governing equations describing both forward and reverse circulation can be described by the pair of equations

$$\delta \frac{dT_p}{dz} = \frac{Q_p - V_p F_p}{\rho c V_p} \tag{80}$$

for flow in the pipe (denoted by the subscript "p"), and

$$-\delta \frac{dT_a}{dz} = \frac{Q_a - V_a F_a}{\rho c V_a} \tag{81}$$

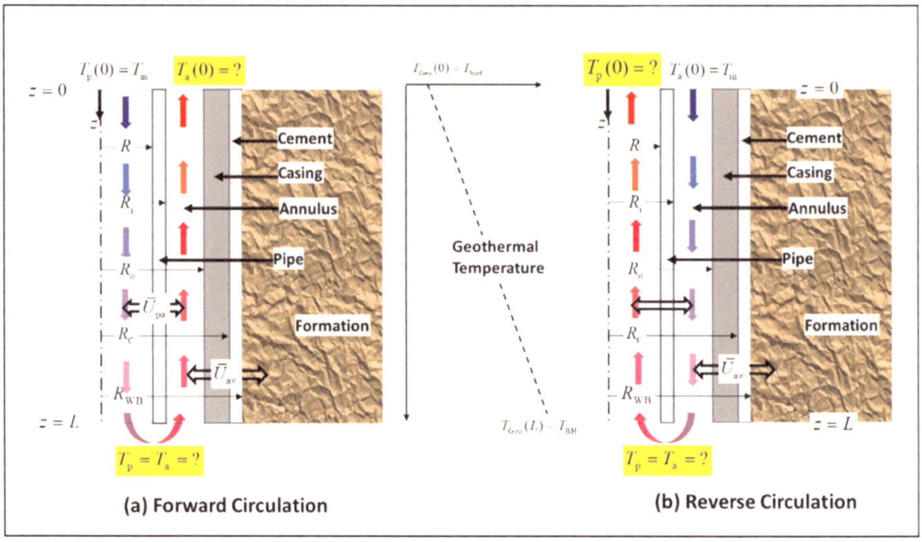

Figure 11.
Thermal interaction in forward (left) and reverse (right) circulation scenarios.

for flow in the annulus (denoted by the subscript "a"). Note that the subscript "p" has been dropped from the specific heat since it is now used to denote the flowing stream in the inner pipe. The direction of the circulation is characterised by the parameter $\delta = \pm 1$, with the positive and negative signs denoting Forward and Reverse circulation, respectively. The associated boundary conditions are

$$(1+\delta)T_{\mathrm{p}}(0) + (1-\delta)T_{\mathrm{a}}(0) = 2T_{\mathrm{in}} \tag{82}$$

at the surface inlet and the matching condition

$$T_{\mathrm{a}}(L) = T_{\mathrm{p}}(L) \tag{83}$$

at the bottom of the well assuming no losses as the fluid leaves one conduit and enters another. The velocities are related through mass conservation such that

$$\dot{m} = \rho A_{\mathrm{p}} V_{\mathrm{p}} = \rho A_{\mathrm{a}} V_{\mathrm{a}} \Rightarrow V_{\mathrm{a}} = \left(\frac{A_{\mathrm{p}}}{A_{\mathrm{a}}}\right) V_{\mathrm{p}} \tag{84}$$

The Heat Transfer terms in Eqs. (80) and (81) are

$$Q_{\mathrm{p}} = 2R^{-2}\overline{U}_{\mathrm{pa}}(T_{\mathrm{a}} - T_{\mathrm{p}}) \tag{85}$$

for the pipe describing the interaction between the two flowing streams and

$$Q_{\mathrm{a}} = 2R^{-2}\overline{U}_{\mathrm{pa}}(T_{\mathrm{p}} - T_{\mathrm{a}}) + 2R^{-2}\overline{U}_{\mathrm{a}\infty}(T_{\mathrm{Geo}} - T_{\mathrm{a}}) \tag{86}$$

for the annulus describing the interaction between the streams and the interaction between the annulus stream and the formation. In accordance with the formulation of the overall heat transfer conductance, we have with respect to the geometry of **Figure 11** the following expressions

$$\frac{\overline{U}_{\text{pa}}}{k_{\text{fluid}}} = \left[\frac{k_{\text{fluid}}}{hR} + \frac{k_{\text{fluid}}}{k_{\text{steel}}} \ln \frac{R_{\text{i}}}{R} + \frac{k_{\text{fluid}}}{h_{\text{i}}R_{\text{i}}}\right]^{-1} \tag{87}$$

and

$$\frac{\overline{U}_{\text{a}\infty}}{k_{\text{fluid}}} = \left[\frac{k_{\text{fluid}}}{h_{\text{o}}R_{\text{o}}} + \frac{k_{\text{fluid}}}{k_{\text{steel}}} \ln \frac{R_{\text{o}}}{R_{\text{C}}} + \frac{k_{\text{fluid}}}{k_{\text{cement}}} \ln \frac{R_{\text{C}}}{R_{\text{wb}}} \frac{k_{\text{fluid}}}{k_{\text{Geo}}} \frac{1}{F(\tau)}\right]^{-1} \tag{88}$$

where τ is the Fourier number corresponding to the instantaneous snapshot in time at which the temperature profiles correspond to the pseudo steady state. In both expressions above, it should be noted that the division by the constant fluid thermal conductivity (assumed) enables the evaluation of the forced convection Nusselt numbers in terms of known correlations such as the Dittus-Boelter or Sieder-Tate models. The frictional heating terms for the pipe and annulus streams are given by

$$F_{\text{p}} = 2\frac{\tau_{\text{p}}(z)}{R} = \frac{f_{\text{p}}}{4} \frac{\rho V_{\text{p}}^2}{R} \tag{89}$$

and

$$F_{\text{a}} = 2\frac{\tau_{\text{a}}(z)}{R_{\text{o}} - R_{\text{i}}} = \frac{f_{\text{a}}}{4} \frac{\rho V_{\text{a}}^2}{(R_{\text{o}} - R_{\text{i}})} \tag{90}$$

Normalising the wellbore streamwise coordinate by the length as in the previous exercises, and the pipe and annulus temperatures by the surface to well depth temperature difference as before yields the following coupled system of equations

$$\frac{\text{d}}{\text{d}\xi}\begin{bmatrix}\theta_{\text{p}} \\ \theta_{\text{a}}\end{bmatrix} = \begin{bmatrix}-\delta N_{\text{pa}} & \delta N_{\text{pa}} \\ -\delta N_{\text{pa}} & \delta(N_{\text{pa}} + N_{\text{a}\infty})\end{bmatrix}\begin{bmatrix}\theta_{\text{p}} \\ \theta_{\text{a}}\end{bmatrix} + \begin{bmatrix}\delta\Lambda_{\text{p}} \\ -\delta\Lambda_{\text{a}}\end{bmatrix} + \begin{bmatrix}0 \\ -\delta N_{\text{a}\infty}\end{bmatrix}\xi \tag{91}$$

which makes use of the fact that the normalised linear geothermal temperature is $\theta_{\text{Geo}}(\xi) = \xi$. Eq. (91) is subject to the pair of boundary conditions

$$\begin{aligned}&(1+\delta)\theta_{\text{p}}(0) + (1-\delta)\theta_{\text{a}}(0) = 2\theta_{\text{in}} \\ &\theta_{\text{a}}(1) = \theta_{\text{p}}(1)\end{aligned} \tag{92}$$

The governing dimensionless parameters are the Number of Transfer Unit parameters

$$N_{\text{pa}} = 2\pi\frac{\overline{U}_{\text{pa}}L}{\dot{m}c} \qquad N_{\text{a}\infty} = 2\pi\frac{\overline{U}_{\text{a}\infty}L}{\dot{m}c} \tag{93}$$

and the dimensionless frictional heating parameters

$$\Lambda_{\text{p}} = \frac{f_{\text{p}}}{4} \frac{V_{\text{p}}^2}{c\Delta T} \frac{L}{R} \qquad \Lambda_{\text{a}} = \frac{f_{\text{a}}}{4} \frac{V_{\text{a}}^2}{c\Delta T} \frac{L}{R_{\text{o}} - R_{\text{i}}} \tag{94}$$

The analytical solution of Eq. (91) yields the pair of equations for the pipe and annulus temperature profiles as

$$\theta_p(\xi) = C_\lambda e^{\lambda\xi} + C_\mu e^{\mu\xi} + h_p + \xi \tag{95}$$

and

$$\theta_a(\xi) = C_\lambda r_\lambda e^{\lambda\xi} + C_\mu r_\mu e^{\mu\xi} + h_a + \xi \tag{96}$$

where λ and μ are the eigenvalues of the matrix in Eq. (91) given by

$$\begin{aligned} \lambda &= \frac{1}{2}\left[\delta N_{a\infty} + \sqrt{N_{a\infty}{}^2 + 4N_{pa}N_{a\infty}}\right] \\ \mu &= \frac{1}{2}\left[\delta N_{a\infty} - \sqrt{N_{a\infty}{}^2 + 4N_{pa}N_{a\infty}}\right] \end{aligned} \tag{97}$$

The r and h constants in Eqs. (95) and (96) are

$$\begin{aligned} r_\lambda &= 1 + \lambda\left(\delta N_{pa}\right)^{-1} \\ r_\mu &= 1 + \mu\left(\delta N_{pa}\right)^{-1} \end{aligned} \tag{98}$$

and

$$\begin{aligned} h_p &= \left(N_{pa}N_{a\infty}\right)^{-1}\left[\left(N_{pa} + N_{a\infty}\right)\Lambda_p + N_{pa}\Lambda_a - \delta N_{a\infty}\right] \\ h_a &= N_{a\infty}{}^{-1}\left(\Lambda_p + \Lambda_a\right) \end{aligned} \tag{99}$$

The constants of integration are determined from the boundary conditions as

$$\begin{aligned} C_\lambda &= \frac{\left(e^\mu - r_\mu e^\mu\right)\left(2\theta_{in} - (1+\delta)p_p - (1-\delta)p_a\right) - \left(p_a - p_p\right)\left[(1+\delta) + (1-\delta)r_\mu\right]}{\left[(1+\delta) + (1-\delta)r_\lambda\right]\left(e^\mu - r_\mu e^\mu\right) - \left[(1+\delta) + (1-\delta)r_\mu\right]\left(r_\lambda e^\lambda - e^\lambda\right)} \\ C_\mu &= \frac{\left(r_\lambda e^\lambda - e^\lambda\right)\left(2\theta_{in} - (1+\delta)p_p - (1-\delta)p_a\right) + \left(p_a - p_p\right)\left[(1+\delta) + (1-\delta)r_\lambda\right]}{\left[(1+\delta) + (1-\delta)r_\lambda\right]\left(e^\mu - r_\mu e^\mu\right) - \left[(1+\delta) + (1-\delta)r_\mu\right]\left(r_\lambda e^\lambda - e^\lambda\right)} \end{aligned} \tag{100}$$

6.1 Forced convection in the annulus

In Eqs. (87) and (88), the terms h_i and h_o represent the heat transfer coefficients at the inner and outer surfaces of the annulus, respectively. In turbulent flows in annuli with radius ratios approaching unity, the following approximation can be used

$$h_i = h_o = \overline{h} = \frac{k_{fluid}}{\overline{D}_{hyd}}\mathrm{Nu_T} = \frac{k_{fluid}}{\overline{D}_{hyd}}C\,\mathrm{Re}^m\mathrm{Pr}^n \tag{101}$$

where $\overline{D}_{hyd}$ is the hydraulic diameter of the annulus, and C, m, and n are the constants of the forced convection correlation used.[10] Most annular flows in wellbore circulating scenarios however, tend to be laminar, and associated with annulus radius

[10] For the very common Dittus-Boelter correlation, the values are $C = 0.023$, $m = 0.8$, and $n = 0.33$.

ratios often well less than unity. In addition, the Non-Newtonian nature of the flow must be considered. Merely replacing the turbulent Nusselt Number Nu_T in Eq. (101) with its laminar analogue Nu_L is not consistent with the physics of the problem. It is recommended that for fluids that obey the Power-Law model, the following correlations from Chandrasekhar [17] be used instead

$$\begin{aligned} \frac{h_i\overline{D}}{k_{\text{fluid}}}(n,\kappa) &= -\frac{2}{1-\theta_b(n,\kappa)}\left(\frac{1-\kappa}{\kappa\ln\kappa}\right) \\ \frac{h_o\overline{D}}{k_{\text{fluid}}}(n,\kappa) &= -\frac{2}{\theta_b(n,\kappa)}\left(\frac{1-\kappa}{\ln\kappa}\right) \end{aligned} \tag{102}$$

where the dimensionless bulk temperature is a function of the power law index and radius ratio, and is given by

$$\begin{aligned} &\theta_b(n,\kappa) = \sum_{j=1}^{4}\left(a_j + nb_j\right)\kappa^{j-1} \\ &a_1 = 0.213, \quad a_2 = 0.576, \quad a_3 = -0.439, \quad a_4 = 0.152 \\ &b_1 = 0.0043, \quad b_2 = -0.0183, \quad b_3 = 0.0236, \quad b_4 = -0.0102 \end{aligned} \tag{103}$$

6.2 Examples of forward and reverse circulation

For a given set of operational parameters, the intermediate calculations and evaluation of the dimensionless parameters is shown in **Figure 12** which represents a case of forward circulation of an oil-based fluid in a fairly typical drilling scenario.

The circulating temperature profiles in the Drillpipe and its annulus are shown in **Figure 13** for the parameters in **Figure 12**, but three different flowrates. It is seen that

Wellbore Geometry	
Well Length (m)	5000
Tubing ID (in.)	4.480
Tubing OD (in.)	5.500
Casing ID (in.)	8.535
Casing OD (in.)	9.625
Hole Dia (in.)	12.250

Circulating Parameters	
Circulation Mode	FWD
Circ. Flowrate (GPM)	300
Inlet Temperature (°C)	50
Surface Temperature (°C)	50
Bottomhole Temperature (°C)	125
Pipe Conductivity (W/m-K)	26.200
Operating Time (hours)	1000

Fluid Specification	
Fluid Type	OBM
Specific Gravity	1.5
PV (cp)	15
YP (lbf/100 ft²)	7

Formation Properties	
Rock Thermal Conductivity (BTU/ft-hr-F)	1.73
Specific Heat (BTU/lbm-F)	0.300
Density (lbm/ft³)	140

Fluid Thermophysical Properties	
Consistency Parameter, K (lbf-sⁿ/ft²)	2.19E-03
Power Law Index, *n*	0.750
Density (lbm/ft³)	93.6
Specific Heat (BTU/lbm-F)	0.295
Thermal Conductivity (BTU/ft-hr-F)	0.488
Thermal Diffusivity (ft2/s)	4.90E-06

Flow Hydraulics	
Tubing Flow Area (ft^2)	0.109
Annulus Flow Area (ft^2)	0.232
Tubing Hydraulic Dia (ft)	0.373
Annulus Hydraulic Dia (ft)	0.253
Volumetric Flowrate (f3/s)	0.668
Mass Flow Rate (lbm/hr)	225,211
Tubing Velocity (fps)	6.106
Annulus Velocity (fps)	2.877

Pseudo-Steady State Parameters	
Thermal Diffusivity (ft^2/hr)	4.13E-02
Fourier Number	158.48
Ramey Function	0.42388

Effective Kinematic Viscosity (f2/s)	
Tubing	0.0103
Annulus	0.0078

Prandtl Numbers	
Tubing	2,099
Annulus	1,599

Reynolds Numbers	
Tubing	221.6
Annulus	92.9

Nusselt Numbers	
Tubing	4.5
Annulus ID	4.5
Annulus OD	3.7

Overall Heat Transfer Conductances	
Tubing-to-Annulus (BTU/ft-hr-F)	0.547
Annulus-to-Formation (BTU/ft-hr-F)	0.344

Friction factors	
Tubing	0.289
Annulus	0.689

Dimensionless Groups	
δ	1
θ_{in}	0.000
N_{pa}	0.847
N_{geo}	0.532
Λ_p	0.237
Λ_a	0.185

Figure 12.
Estimation of dimension groups from problem data (forward circulation).

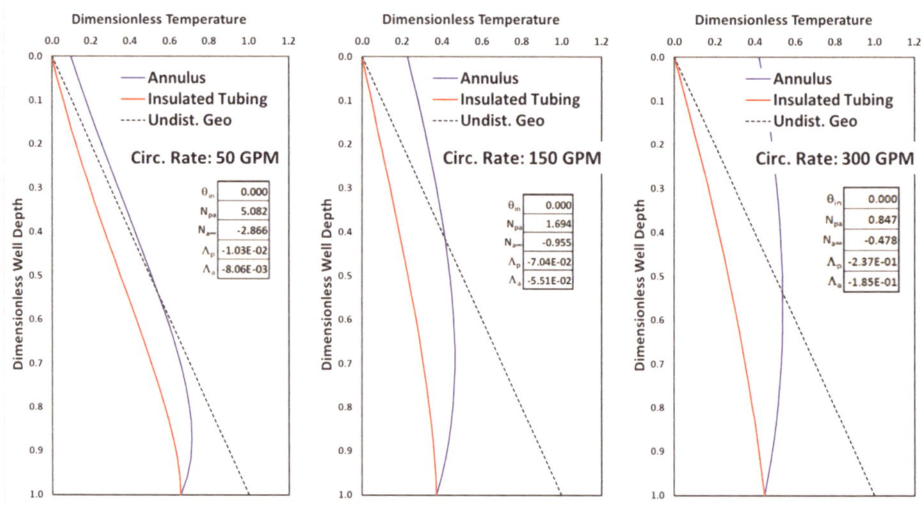

Figure 13.
Circulating temperature profiles for three different flowrates – Forward circulation of an oil-based drilling fluid.

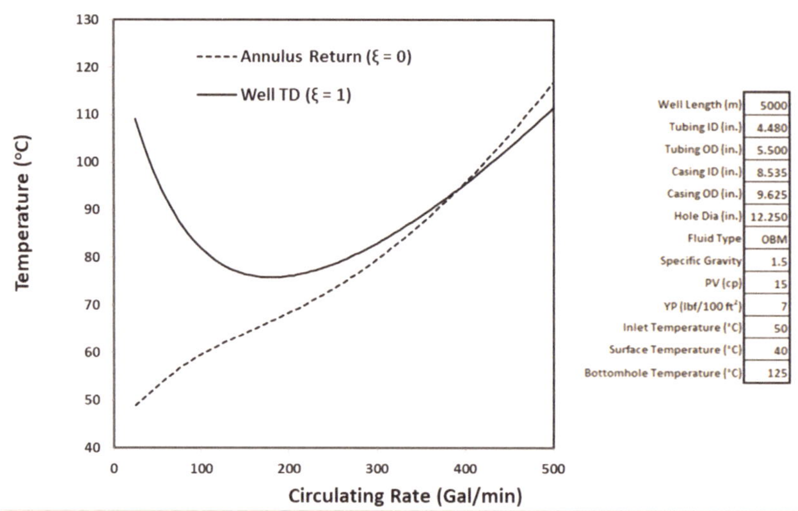

Figure 14.
Negative joule-Thomson (frictional heating) effect of circulating flow rate on well temperatures.

the temperature at the well TD at first decreases with flowrate as would be expected, but at higher flowrates, actually increases due to frictional heating. This is more clearly evident in **Figure 14** where the TD and arrival temperatures are plotted over a range of flowrates. The inflexion point in the well TD temperature is where the negative Joule-Thomson effect surpasses the advection effect in the drillpipe. The point at which the (dashed) arrival temperature curve intersects the (solid) well TD curve is where frictional heating is significant even in the annulus.

The temperature profiles shown in **Figure 15** for three different mass flow rates of water in a reverse circulating scenario correspond to a geothermal well. In this case, the inner conduit known as the tubing is assumed to be insulated as is common in

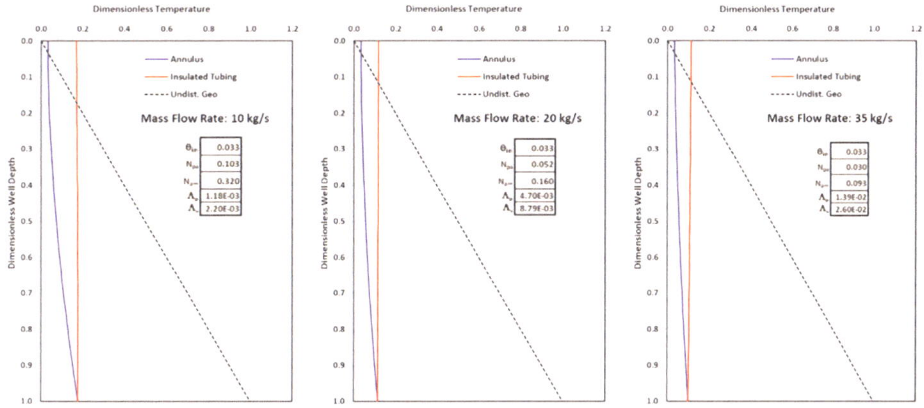

Figure 15.
Circulating temperature profiles for three different flowrates – Reverse circulation of water in a geothermal well with an insulated tubing.

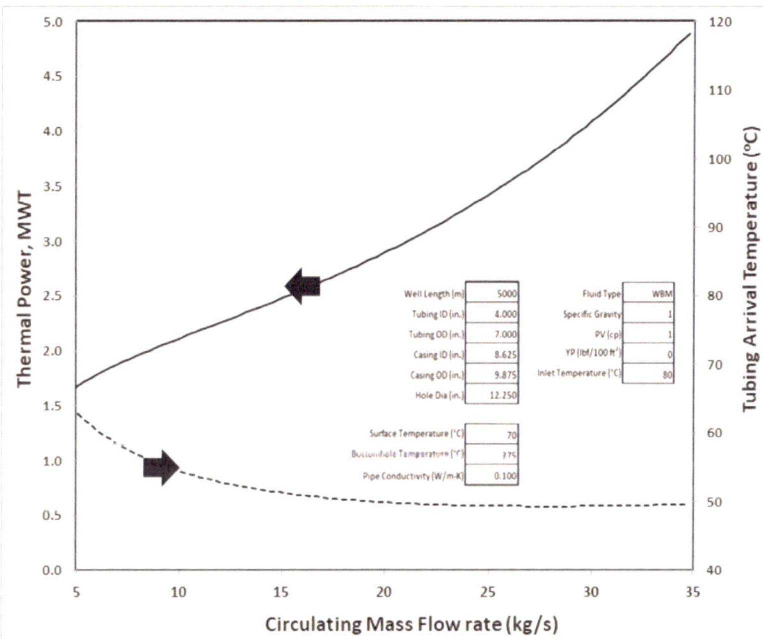

Figure 16.
Effect of flow rate on thermal power generated and arrival temperature in a geothermal well with an insulated tubing.

geothermal wells. The well TD and arrival temperatures quickly tend to become independent of mass flow rate. The thermal power produced by a geothermal well is given by

$$\dot{P}_{\mathrm{MWT}} = \dot{m} c_{\mathrm{P}} (T_{\mathrm{arr}} - T_{\mathrm{inlet}}) \tag{104}$$

and is plotted as a function of mass flow rate as shown in **Figure 16**. The arrival temperature is seen to become independent of mass flow rate at about 25 kg/sec whereupon the thermal power increases linearly. It is important to note that only a fraction of the thermal power is actually converted into electric power (often at a rate of about 15–20%) that can be transmitted to a grid. Furthermore, a portion of even this converted power has to be used to overcome the parasitic power due to frictional losses in the geothermal wellbore. The thermal power serves however as a useful metric in a parametric sensitivity analysis of a geothermal well.

The circulating thermal model developed here considers only a simple monobore well for illustrative purposes. For an extension of the methodology to a complex wellbore with multiple wellbore segments coupled via an arbitrary number of interface temperature matching conditions, the reader is referred to [9]. That study also considers curvature and tortuosity effects in deviated wells, variable lithology, multiple geothermal gradients, and the effects of fluid rheology.

7. Evaluation of wellbore and interface temperatures

7.1 Wellbore temperatures in a transient analysis

The methodologies described in the previous sections dealt with the estimation of flowing temperatures in operating and circulating scenarios. An issue of equal - if not often greater - importance is the estimation of temperatures in the intervening wellbore layers (fluid and solid) between the flow conduit and the formation. The description of a fully transient analysis wherein the transient temperatures in all layers are evaluated in tandem with the transient flowing temperature is beyond the scope of this chapter, but the interested reader is referred to [3] where such an analysis is described in near-exhaustive granularity. What will be demonstrated in what follows here is how to estimate the interface thermal conductivities and temperatures which are needed to evaluate the heat fluxes as well as the natural convection multipliers required for the estimation of the nodal thermal conductivities in fluid layers.

Consider the depiction in **Figure 17** showing 3 adjacent layers designated $i-1$, i and $i+1$ from left to right. The barred and unbarred symbols refer to interfacial and nodal entities respectively. At the interface $i-1$ at the left of the layer i, the flux expressed in terms of the straddling nodal difference and the interface conductivity, can also be expressed in terms of the differences between the nodes and the interface, and the nodal conductivities. This relationship can be expressed as

$$\dot{q}_{i-1}(z) = -\frac{\overline{k}_{i-1}}{\ln {}^{r_i}\!/_{r_{i-1}}}(T_{i-1} - T_i) = -\frac{k_{i-1}}{\ln {}^{\overline{r}_{i-1}}\!/_{r_{i-1}}}(T_{i-1} - \overline{T}_{i-1}) = -\frac{k_i}{\ln {}^{r_i}\!/_{\overline{r}_{i-1}}}(\overline{T}_{i-1} - T_i) \quad (105)$$

from which the interface thermal conductivity at the right and left interfaces with indices i and $i-1$ can be expressed in terms of the nodal values as the weighted harmonic means

$$\begin{aligned} \overline{k}_i &= \ln {}^{r_{i+1}}\!/_{r_i}\left(\frac{\lambda_i}{k_i} + \frac{\mu_i}{k_{i+1}}\right)^{-1} \\ \overline{k}_{i-1} &= \ln {}^{r_i}\!/_{r_{i-1}}\left(\frac{\lambda_{i-1}}{k_{i-1}} + \frac{\mu_{i-1}}{k_i}\right)^{-1} \end{aligned} \quad (106)$$

where the geometric coefficients are

$$\lambda_i = \ln \bar{r}_i/r_i \qquad \mu_i = \ln r_{i+1}/\bar{r}_i \tag{107}$$

The interface temperatures can be expressed in terms of the geometric coefficients in Eq. (107) and the nodal conductivities as

$$\begin{aligned} \overline{T}_i &= \frac{\lambda_i^{-1} k_i T_i + \mu_i^{-1} k_{i+1} T_{i+1}}{\lambda_i^{-1} k_i + \mu_i^{-1} k_{i+1}} \\ \overline{T}_{i-1} &= \frac{\lambda_{i-1}^{-1} k_{i-1} T_{i-1} + \mu_{i-1}^{-1} k_i T_i}{\lambda_{i-1}^{-1} k_{i-1} + \mu_{i-1}^{-1} k_i} \end{aligned} \tag{108}$$

When there is at least one fluid layer subject to natural convection, the evaluation of the interface values must be embedded in an iterative sequence within each time step. Note that in a transient analysis, the fluxes on either side of a nodal layer need not be equal, so that in general, $\overline{q}_{i-1} \neq \overline{q}_i$ with respect to **Figure 17**.

7.2 Wellbore temperatures in a Pseudo steady state analysis

If the transients are relegated solely to the formation and included as a flux captured at a snapshot in time, then the overall heat transfer coefficient can be calculated from Eq. (32) without any need for explicitly formulating an energy balance for each individual wellbore layer. If at least one layer is a fluid layer, then the interface temperatures are required to model the natural convection which then renders the procedure iterative. At each step of the iteration, the interface temperatures can be calculated as follows - given a value of the fluid temperature evaluated a certain iteration step, the temperature at the conduit wall is estimated from the forced convection component of the overall thermal resistance as

$$\overline{T}_0 = T_0 - \frac{\overline{U}}{hR}\left[T(z) - T_{\text{geo}}(z)\right] \tag{109}$$

Subsequently the temperatures at each of the outer layers is evaluated as

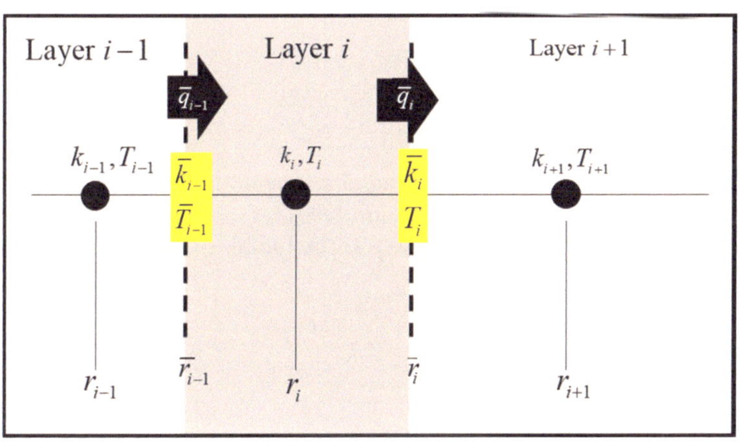

Figure 17.
Nodal and Interface temperatures and conductivities.

$$\overline{T}_j = \overline{T}_{j-1} - \ln\left(\frac{\overline{r}_j}{\overline{r}_{j-1}}\right)\frac{\overline{U}}{k_j}\left[T(z) - T_{\text{geo}}(z)\right], \quad j = 1\cdots N \tag{110}$$

Once the iteration has converged, the temperature at the layer mid radius is evaluated as

$$T_j = \frac{1}{2}\left(\overline{T}_{j-1} + \overline{T}_j\right), \quad j = 1\cdots N \tag{111}$$

Note that in this case, since the fluxes are evaluated using interface temperature differences, there is no need to evaluate interface thermal conductivities as in the transient analysis case.

8. Conclusions

The key aspects of wellbore heat transfer cover the entire gamut of thermal energy transport mechanisms from advection/convection in wellbores, to conduction across wellbore tubular and cement layers, to natural convection in trapped annuli, and diffusion in semi-infinite domains from a wellbore to the surrounding formation layer. A term by term derivation of the transport equation using the enthalpy formulation is crucial to understanding the relative importance of the various energy terms.

The mathematical models developed are applicable to a very wide range of wellbore operations, from production and injection to circulation. While thermal transients can generally be ignored for long term production scenarios, significant errors can result from ignoring them in shorter injection and circulating scenarios. When the flowrates exceed a certain threshold, the seemingly counter-intuitive temperature profiles can be explained in terms of the Negative-Joule-Thomson effect.

The most efficient approach to solving transient wellbore heat transfer problems is by Laplace transformation of the governing equations. An efficient method of inversion back to the physical domain is with the use of the powerful Gaver-Stehfest function sampling algorithm.

Appendix

Fourier-Bessel coefficients

The solution to the one dimensional radial transient diffusion problem defined by Eq. (61) with a constant boundary condition ($\phi(1, \xi, \tau) = 1$ in Eq. (63)) at the inner boundary and an insulated outer boundary at the radial location η_∞ is

$$\phi(\xi, \tau) = 1 - \sum_{j=1}^{\infty} C_j F_j(\eta)\, e^{-\lambda_j^2 \tau} \tag{112}$$

in which the radial eigenfunction is

$$F_k(\eta) = J_1(\mu_k \eta_\infty) Y_0(\mu_k \eta) - Y_1(\mu_k \eta_\infty) J_0(\mu_k \eta) \tag{113}$$

where $J_k()$ and $Y_k()$ for $k = 0, 1$ are the Bessel Functions of the First and Second Kind of Order k and the eigenvalues are the zeroes of

$$J_1(\mu\eta_\infty)Y_0(\mu) - Y_1(\mu\eta_\infty)J_0(\mu) = 0 \tag{114}$$

The Fourier-Bessel coefficients are

$$C_k = \left[\int_1^D [F_k(\eta)]^2 \eta \mathrm{d}\eta\right]^{-1} \int_1^D F_k(\eta)\eta \mathrm{d}\eta \tag{115}$$

The flux at the inner boundary is the entity of interest, so that the coefficient $\overline{C}_j$ in Eq. (65) is given by

$$\overline{C}_j = C_j \frac{\partial F_j}{\partial \eta}\bigg|_{\eta=1} = C_j[Y_1(\mu_k\eta_\infty)J_1(\mu_k) - J_1(\mu_k\eta_\infty)Y_1(\mu_k)] \tag{116}$$

Stehfest accelerators

The coefficients in the Gaver Summation known as Stehfest accelerators are defined as

$$\lambda_i = (-1)^{\mathrm{Int}\left(\frac{N}{2}+i\right)} \sum_{k=\mathrm{Int}\left(\frac{i+1}{2}\right)}^{k=\mathrm{Min}\left(i,\frac{N_{GS}}{2}\right)} \frac{k^{\frac{N_{GS}}{2}}(2k)!}{\left(\frac{n}{2}-k\right)!k!(k-1)!(i-k)(2k-i)!}, k = 1\cdots N_{GS} \tag{117}$$

and are tabulated below for the first few even orders of the method.

k	Stehfest accelerators				
	2	4	6	8	10
1	2	−2	1	−3.333E−01	8.333E−02
2	−2	26	−49	4.833E+01	−3.208E+01
3		−48	366	−9.060E+02	1.279E+03
4		24	−858	5.465E+03	−1.562E+04
5			810	−1.438E+04	8.424E+04
6			−270	1.873E+04	−2.370E+05
7				−1.195E+04	3.759E+05
8				2.987E+03	−3.401E+05
9					1.641E+05
10					−3.281E+04

Table 1.
Stehfest accelerators.

Nomenclature

A	Flow Area (m^2)
c_{JT}	Joule-Thomson Coefficient (K-m^2/N)
c_p	Specific Heat at Constant Pressure (J/kg-K)
$\overline{D}$	Hydraulic Diameter (m)
e	Total Specific Energy (J/kg-K)
f	Friction Factor
g	Acceleration due to Gravity (m^2/s)
h	Specific Enthalpy (J/kg) / heat transfer coefficient (W/m^2-K)
k	Thermal Conductivity (W/m-K)
L	Conduit Length (m)
$\dot{m}$	Mass Flow Rate (kg/s)
n	Power Law Index (dimensionless)
Nu	Nusselt Number (dimensionless)
Nu	Nusselt Number (dimensionless)
P	Pressure (N/m^2)
Pr	Prandtl Number
Pr	Prandtl Number (dimensionless)
Re	Reynolds Number (dimensionless)
$\dot{q}$	Heat Fux per Unit Area (W/m^2)
r	local radius (m)
$\bar{r}$	Interface Radius (m)
R	Conduit Radius (m)
$\overline{R}$	Wellbore Outer Radius (m)
Re	Reynolds Number (dimensionless)
Re	Reynolds Number (dimensionless)
s	Laplace variable in Frequency Domain
t	time (s)
T	Temperature (K)
u	Specific Internal Energy (J/kg)
U	Overall Heat Transfer Coefficient (W/m^2-K)
$\overline{U}$	Overall Heat Transfer Conductance (W/m-K)
V	Flow Velocity (m/s)
y	True Vertical Depth (m)
z	Streamwise Coordinate (m)
α	Isobaric Volume Expansivity(1/K)
β	Isothermal Compressibility (m^2/N)
κ	Annulus Radius Ratio ()
δ	Directional Index
ϕ	Dimensionless Formation Temperature
η	Dimensionless radial Coordinate
ν	Kinematic Viscosity (m^2/s)
θ	Dimensionless Fluid Temperature/ Wellbore Local Inclination Inclination
ρ	Fluid Density (kg/m^3)
τ	Dimensionless Time (Fourier Number)
τ_w	Wall Spear Stress (Pa)
ξ	Dimensionless Streamwise coordinate
a	Annulus

Form	Formation
Geo	Geothermal
i	Inner Surface
o	Outer Surface
p	Pipe
wb	Wellbore

Author details

Sharat V. Chandrasekhar*, Udaya B. Sathuvalli and Poodipeddi V. Suryanarayana
Blade Energy Partners, Frisco, TX, USA

*Address all correspondence to: schandrasekhar@blade-energy.com

References

[1] Lesem LB, Greytok F, Marotta F, McKetta JJ. A method of calculating the distribution of temperature in flowing gas wells. Transactions on AIME. 1957; **210**:169

[2] Moss JT, White PD. How to calculate temperature profiles in a water-injection well. Oil and Gas Journal. 1959;**57**(11):174

[3] Chandrasekar S. Revisiting the classical problem of transient temperature prediction in complex Deepwater wellbores with a new semi-analytical approach. In: SPE/IADC Paper 199653. Galveston, TX; 2020

[4] Ramey HJ. Wellbore heat transmission. In: SPE Paper 96. 1962

[5] Wilhite GP. Over-all heat transfer coefficients in steam and hot water injection wells. In: SPE Paper 1449. Denver, CO; 1966

[6] Raymond L. Temperature distribution in a circulating drilling fluid. In: SPE Paper 2320. Austin, TX; 1969

[7] Wooley GR. Computing downhole temperatures in circulation, injection and production wells. Journal of Petroleum Technology. 1980;**1980**: 1509-1522

[8] Wu Y-S, Pruess K. An analytical solution for wellbore heat transmission in layered formations. SPE Reservoir Engineering Journal. 1990;**1990**:531-538

[9] Chandrasekhar S, Sathuvalli UB, Suryanarayana PV. Circulating temperatures in complex wellbores: A quasi-exact solution. In: SPE-IADC Bangkok paper 191033. 2018

[10] Hasan AR, Kabir CS. Fluid Flow and Heat Transfer. 1st ed. Richardson, TX: Society of Petroleum Engineers; 2002

[11] Mitchell RF, Sathuvalli UB. Chapter 10 in Advanced drilling and well technology. In: Heat Transfer in Wells – Wellbore Thermal and Flow Simulation. Richardson, TX: SPE; 2009

[12] Jones RR. A novel economical approach for accurate real-time measurement of wellbore temperatures. In: SPE-ATCE Paper 15577. New Orleans, LA; 1986

[13] Aeschliman DP, Meldau RF, Noble NJ. Thermal efficiency of a steam injection test well with insulated tubing. In: SPE Paper 11735. Ventura, CA; 1983

[14] Dropkin D, Sommerscales E. Heat transfer by natural convection in liquids confined by two parallel plates which are inclined at various angles with respect to the horizontal. Journal of Heat Transfer. 1965;**87**(1):77-82

[15] Ozisik MN. Heat Conduction. 2nd ed. Hoboken, NJ: John Wiley & Sons; 1993

[16] Stehfest H. Algorithm 368: Numerical inversion of Laplace transforms. Communications of the ACM. 1970;**13**(1):47-49

[17] Chandrasekhar S. Annular Couette–Poiseuille flow and heat transfer of a power-law fluid – analytical solutions. Journal of Non-Newtonian Fluid Mechanics. 2020;**286**

Chapter 6

Heat Transportation by Acicular Micro-Textured Device with Semi-Regular Alignment

Tatsuhiko Aizawa, Hiroki Nakata and Takeshi Nasu

Abstract

Heat transportation device was developed to improve the cooling capacity through the heat convection process and to make low-temperature radiation from the heat source to the objective body in vacuum. This device consisted of the metallic substrate and the acicular micro−/nano-textures in semi-regular alignment. The micro-cone unit cell size and pitch in these textures was controllable by tuning the total current and the current density in the electrochemical processing. Four devices with various unit cell sizes and pitches were prepared for geometric characterization by SEM (Scanning Electron Miscopy) and for spectroscopic analyses on the IR-emittance by FT-IR (Fourier Transform-InfraRed) spectroscopy. Heat radiation experiment was performed to describe the heat transportation in vacuum from the heat source at 323 K to the objective plate. The texture size effect on the low-temperature heat radiation was investigated to build up a physical model for this heat radiation device. Heat convection experiment was also performed to describe the cooling capacity of device under the forced air flow. The unit cell height effect on the cooling behavior was discussed to deduce the physical model for this heat convection device. These models were considered to be used in the computational fluid mechanics simulations.

Keywords: acicular micro−/nano-texturing, heat transportation device, size effect, FT-IR diagnosis, resonance, IR-emission, heat radiation, heat convection, cooling capacity, theoretical model

1. Introduction

Heat transfer and transportation [1–3] is an essential engineering to design the power generator, engines, reactors, and electric devices, to describe the thermodynamic behavior in those systems and units and to develop nontraditional methods for energy saving, reduction of thermal waste, and efficient thermal management. In a sustainable society or in a circular economy toward the zero-carbon wastes and carbon neutrality [4], various mechanical and electrical systems must be designed to significantly minimize the energy consumption and thermal waste, to prevent them from the thermal damages and to prolong their lifetime and make full use of generated heat.

As discussed in [5–8], the graphene solid with higher thermal conductivity than copper was utilized as a thermal spreader from the GaN chips to the substrate in order to preserve the integrity of semiconductor packages. In this heat transportation design, higher thermal conductivity than 1000 W/K was fully utilized for fast and homogeneous thermal spreading to prevent the GaN-chipped package from thermal damages. Let us consider how to improve the heat transfer and transportation capacity in three schemes in **Figure 1**, except for the heat transfer from solid to solid.

The boiling heat transfer in **Figure 1a** advances with the phase transformation from liquid to vapor; its essential behavior is described by the boiling curve. Various approaches [9–11] were proposed to control this relationship between the heat flux (q) and the superheat (ΔT). Among them, the concave and convex micro-textured interface between the copper heater and the cooling water worked to reduce the onset superheat to start the vapor nucleation, to increase the heat penetration rate, and to attain the higher heat flux than the burnout critical limit [12–14].

In the heat convection transfer in **Figure 1b**, the solid interface is cooled down by the natural or forced flow of liquid or gaseous media. Extension of the interfacial area improves the overall heat penetration factor through the interface [15, 16]. In addition, the acicular micro-textured interface is cooled more rapidly than that on the bare copper sheet as stated in [17]. The topological design on the micro-textured interface has large influence on the local convection heat transfer to the coolants.

The heat radiation from the solid to the gaseous media or to the vacuum in **Figure 1c** is driven by emission of electromagnetic waves in the IR (Infrared) wavelength range [18]. In particular, the nano-structured, the meta-material, and the concave micro-textured interfaces [19–21] provided a way to make low-temperature heat radiation. As discussed in [17, 22, 23], the heat radiation takes place at 323 K in air and in vacuum from the acicular micro-textured emitter to the objective body. No temperature rose on the object surfaces when using the bare copper substrate without acicular microtextures. As investigated in [24, 25], the wavelength for IR-emission from the textures with semi-regular alignment of micro-cone shaped unit cells was determined by the resonance condition on the micro-cavity, which was surrounded by micro-cone unit cells under high heat flux. This resonation drives the low temperature heat radiation from the textured surface to objectives with the selected IR-wavelength range.

In the present paper, the micro-cone unit cell size in this acicular micro-texturing is varied to investigate the size effect on the IR-emission. The resonance wavelength range for IR-emission is mainly determined by the micro-cone cell height ($H = H_{ave} + H_{dev}$ for the average height, H_{ave}, and the height standard deviation, H_{dev}). This IR-emission is characterized by the resonance condition via $\lambda \sim 2 \times H$.

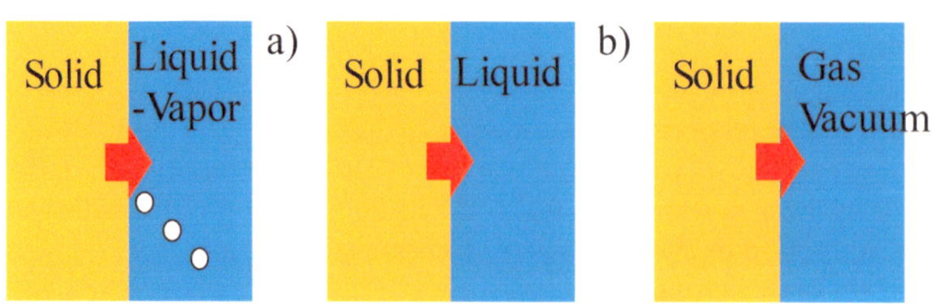

Figure 1.
Three heat transfer and transportation schemes from solids to liquid and gaseous media. (a) Boiling heat transfer, (b) heat convection, and (c) heat radiation.

The low temperature heat radiation experiment is performed to describe the heat transportation from IR-emitter to object in vacuum and to discuss the micro-cone cell size effect on the measured thermal transients on the object. The convection heat transfer under forced airflow is also preformed to understand the cell size effect on the cooling capacity and the cooling power.

2. Methods and materials

An acicular microtexture is formed onto the copper sheet in semi-regular alignment. Its micro-cone shaped unit cell size is controlled to yield several acicular micro-textured specimens. They are characterized by SEM and FT-IR for geometric measurement and emittance of IR (infrared) electromagnetic waves, respectively. Heat convection and radiation experiments are performed to investigate the unitcell size effect on the cooling capacity and on the low-temperature heat radiation behavior.

2.1 Fabrication of acicular micro−/nano-textured substrate

The electrochemical process was utilized to form the acicular microtextures onto the copper sheet. Iron–nickel ionized solution was used to deposit the Fe–Ni alloy microtextures onto the copper sheet. As schematically depicted in **Figure 2**, the nucleation sites for deposition were first built by the femtosecond laser texturing to control the regularity in the alignment of deposits [26, 27]. Then, the wet plating was performed to make fine micro-texturing. To be discussed later, the micro-cone unit cell size is controlled by the total current and current density in this process.

2.2 Control of micro-cone unit cell sizes

Typical three acicular micro-textured specimens were selected to characterize the micro-cone size and regularity in alignment, using their microstructure observation. **Figure 3** depicts the SEM images for LL-, M-, and SS-specimens, respectively.

Each microtexture is geometrically characterized by the micro-cone cell height (H), the root diameter (B), and the mutual distance (D). The regularity in alignment is also controlled by varying the unit cell size. In the following, {H, B, D} for each acicular microtexture is measured from each SEM image.

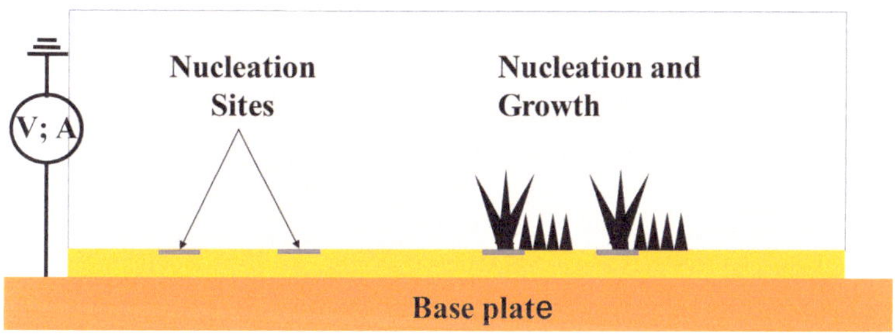

Figure 2.
A schematic view on the electrochemical process with the use of nucleation site control.

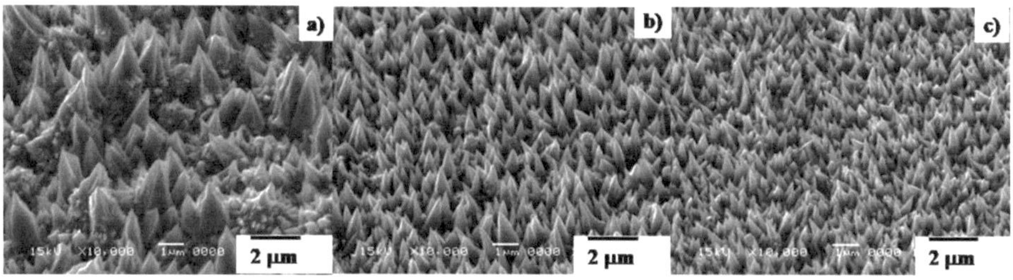

Figure 3.
SEM image on the several micro-textured specimens by varying the unit cell sizes [24, 25]. (a) LL-specimen, (b) M-specimen, and (c) SS-specimen.

2.3 Characterization on the acicular microtextures

Using various SEM images with controlling the skew angle, {H, B, D} are estimated with aid of the picture processing [28] and the computational geometry [29]. On the SEM image in plain view, each micro-cone unit cell is identified as a Voronoi polygon by tessellating every micro-cone unit cell. **Figure 4** shows the tessellation process of SEM image to an assembly of Voronoi polygons. The root diameter (B) of each micro-cone cell is determined by the diameter of an inscribing circle to each Voronoi polygon. The mutual distance (D) is calculated by the distancing length between the center points of neighboring Voronoi polygons. On the SEM image with the skew angle of 30°, the whole micro-cone height profile is digitized in pixels. Using the one-to-one correlation between the measured height and the pixels for the same micro-cone unit, the digitized profile is transformed to the height distribution. Hence, {H, B, D} data are determined for each micro-cone unit cell.

The optical properties of acicular microtextures are measured by the FT-IR (Fourier Transform Infra-Red) spectroscopy with reference to the black body. The FT-IR measurement setup is schematically depicted in **Figure 5a**. The detector was placed onto the specimen to measure the transmittance of electromagnetic waves or T (λ) in

Figure 4.
Geometric characterization on the unit cell size and alignment by picture processing and computer geometry [22, 25].

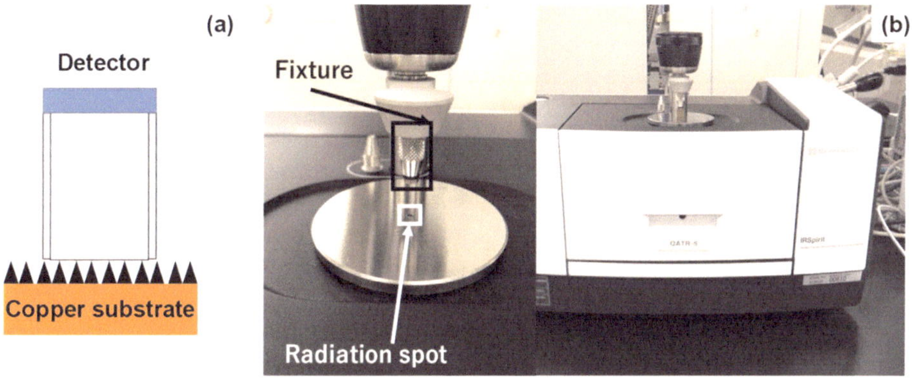

Figure 5.
FT-IR spectroscopy to measure the transmittance spectrum of the micro-textured devices. (a) A schematic view of FT-IR spectroscopic analysis, and (b) overview on the FT-IR spectrometer.

the wavelength (λ) range from the visible light to far-IR. As shown in **Figure 5b**, the FT-IR system (IR Spirit; Shimazu Co., Ltd., Kyoto, Japan) with the detector (QATR-S) was utilized for this spectroscopic measurement. The software (LabSolution IR) was also used to edit the measured raw data and to represent them as a transmittance spectrum, T (Λ), as the function of the wave number (Λ).

2.4 Heat radiation experiment

The heat radiation process advances from the IR-emitter to an objective body even through the vacuum. In this experiment, the test device was placed onto the hot plate with the holding temperature at 323 K. The black-colored polycarbonate (BC-PC) plate with a thickness of 2 mm was used as an objective body. As depicted in **Figure 6a**, thermography (FLIR-100; FLIR Co., Ltd. Frankfurt, Germany) was

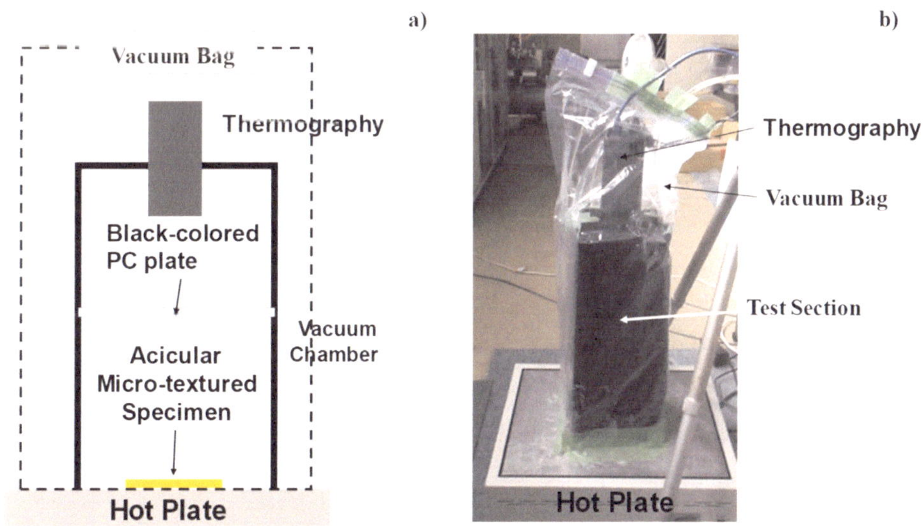

Figure 6.
Low-temperature heat radiation experiment in vacuum from the IR-emitter to the objective solid plate. (a) A schematic view of experimental setup, and (b) overview of the experimental setup.

utilized to measure the thermal transients on the top surface of the BC-PC plate. This thermography was focused on the top surface only to measure the thermal transients on it. The whole setup was fully covered by the black cloth to minimize the optical noise. The test-section setup and thermography were all packed into the vacuum bag for evacuation during the experiment. The pressure in the side of bag was kept to be less than 100 Pa. Throughout the experiment, the vacuum bag was continuously evacuated to be free from the heat convection through the residual air.

In experiment, the heat is first transported to the bottom surface of BC-PC plate *via* the heat radiation process. Then, the temperature rises on its top surface by heat conduction from its bottom to its top surface.

When using the bare copper plate instead of a device, no temperate increase was detected on the top surface of the BC-PC plate. This thermal transient measurement was free from a convection heat transfer below the bottom surface of the BC-PC plate. **Figure 6b** shows the overview of the experiment setup for the measurement of the thermal transients on the top surface of the BC-PC plate.

2.5 Heat convection experiment

An experimental setup for heat convection test under the forced air flow was schematically illustrated in **Figure 7a**. **Figure 7b** shows the overview of the whole setup. The specimen was placed on the hot plate with the holding temperature of 50°C or 323 K. The temperature distribution was monitored by thermography. A bare copper plate was utilized as a reference to investigate the cooling behavior under the forced air flow and the temperature recovery process without the air.

3. Results

SEM was utilized to make geometric diagnosis on the acicular microtextures with aid of the picture processing and the computational geometry. IR-emittance was analyzed by FT-IR spectroscopy to discuss the resonance condition between the peak wavelengths in the emittance spectrum and the micro-cone heights. The thermal transportation from the IR-emitter to the object in vacuum was investigated to demonstrate the heat radiation capacity of acicular microtextured device. The convection heat transfer experiment was performed to describe its higher cooling capacity under the forced air.

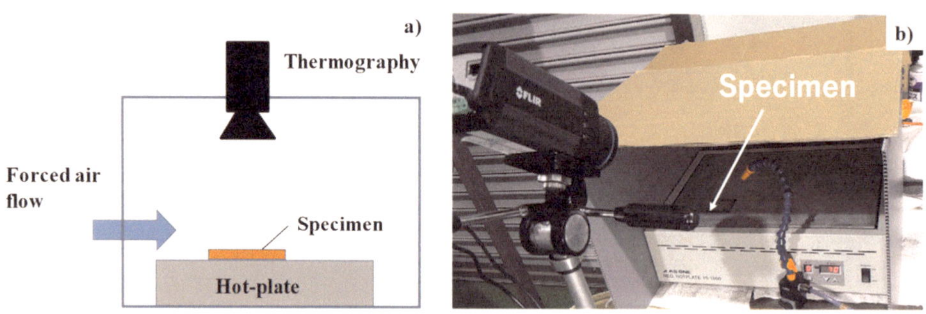

Figure 7.
Heat convection experiment under forced air cooling. (a) A schematic view on the experimental setup, and (b) Its overview.

3.1 Geometric diagnosis on the acicular textured device

Four specimens were prepared to have different micro-cone unit cell sizes for geometric diagnosis. At first, the LL-specimen was utilized to make its geometric diagnosis by measurement of the statistical distributions on {H, B, D}. As shown in **Figure 8a**, the population of micro-cone heights, P (H), has a bimodal distribution with two peaks at H_{ave} = 2.7 μm and 3.5 μm. The standard deviation is nearly the same between the two profiles; ΔH = 0.25 μm. In **Figure 8b**, P(B) also has a bimodal distribution with two medians at 1 μm and 2 μm. The standard deviation of the first profile is ΔB = 0.5 μm, while it becomes 0.75 μm for the second profile. In **Figure 8c**, P(D) has a monotonic distribution with the median at 2.2 μm and the standard deviation of ΔD = 0.75 μm. These P (H), P (B), and P (D) as well as {H_{ave}, B_{ave}, D_{ave}} and {ΔH, ΔB, ΔD} are varied by the regularity control in the alignment of micro-cone unit cells. In the following, P (H), H_{ave}, and ΔH are estimated for other three specimens.

Figure 9 compares the micro-cone unit cell height distributions, P (H), among the M-, S-, and SS-specimens. These P (H) profiles have a mono-modal distribution with the monotonic decrease of H_{ave} to lower heights.

Table 1 summarizes the estimated median, H_{ave}, and standard deviation, ΔH, for these four specimens.

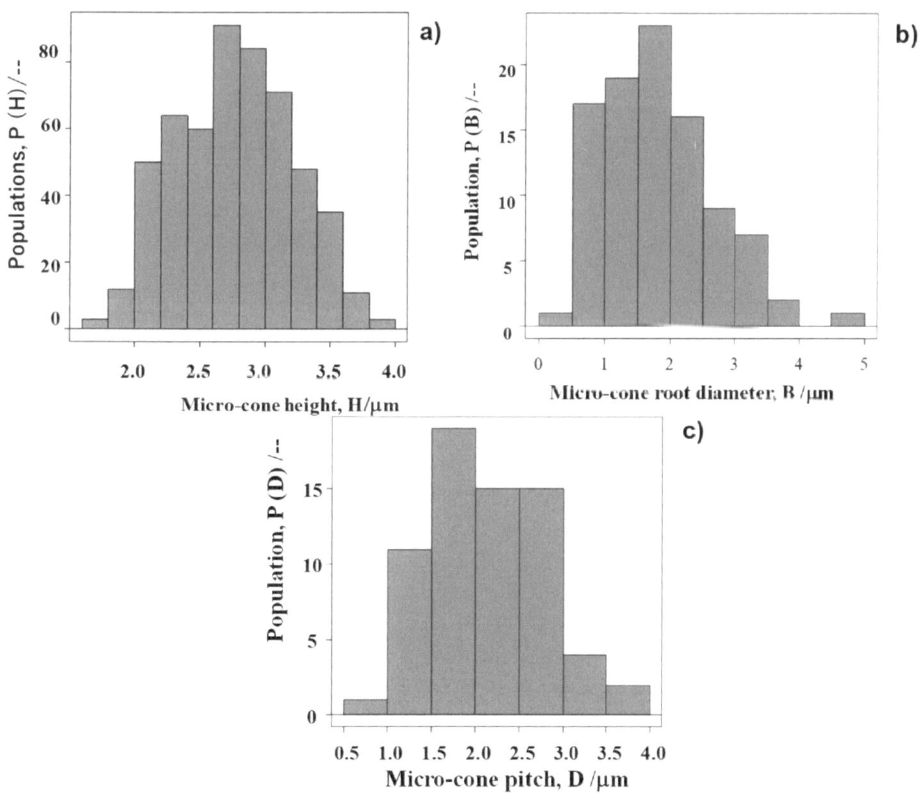

Figure 8.
Populations of {H, B, D} of micro-cone unit cells in the acicular microtextures for the LL-specimen in ***Figure 3a*** *[25]. (a) P (H), (b) P (B), and (c) P (D).*

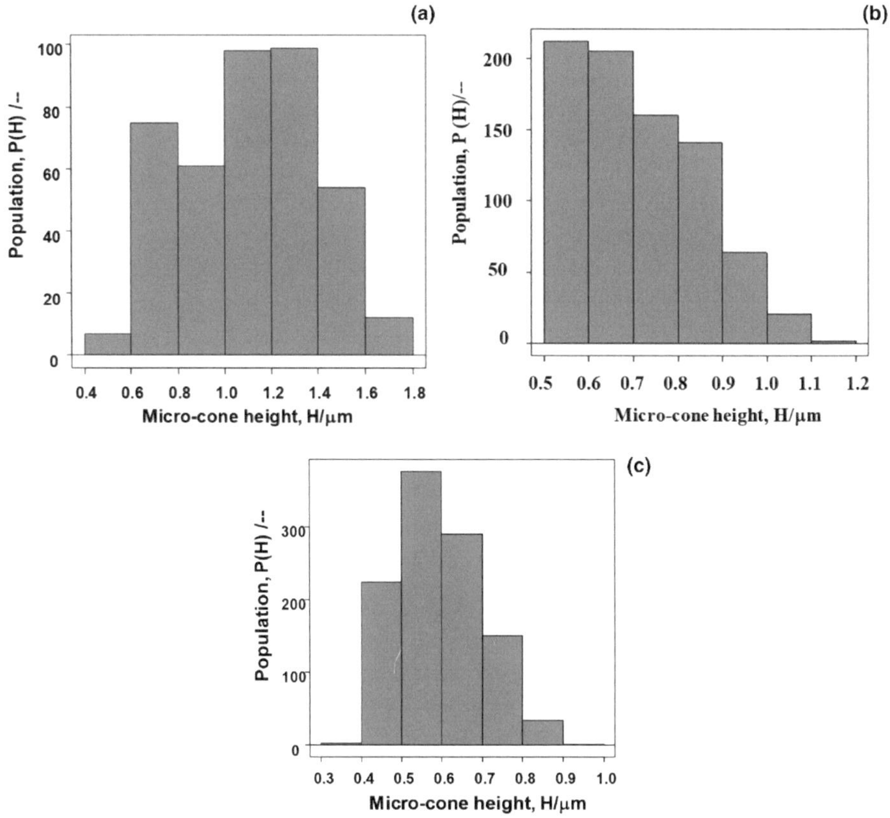

Figure 9.
The micro-cone height distribution, P (H), for M-type, S-type, and SS-type acicular microtextured specimens [25]. (a) P (H) for M-type specimen, (b) P (H) for S-type specimen, and (c) P (H) for SS-type specimen.

Acicular microtextured specimen	H_{ave}, median in P (H) (μm)	ΔH, standard deviation in P (H) (μm)
LL-specimen	3.5	0.25
LL-specimen	2.7	0.25
M-specimen	1.3	0.2
M-specimen	0.8	0.25
S-specimen	0.68	0.12
SS-specimen	0.57	0.1

Table 1.
Median and standard deviation of the micro-cone unit cell height distributions for four specimens.

3.2 Spectroscopic analysis on the IR-emittance from device

The IR-emittance for every acicular micro-textured specimen with the size of 50 mm x 50 mm x 0.5 mmt was estimated from the measured IR-transmittance, T (Λ), by the FT-IR. The measured IR-emittance at the center point of the LL-specimen is shown in **Figure 10**.

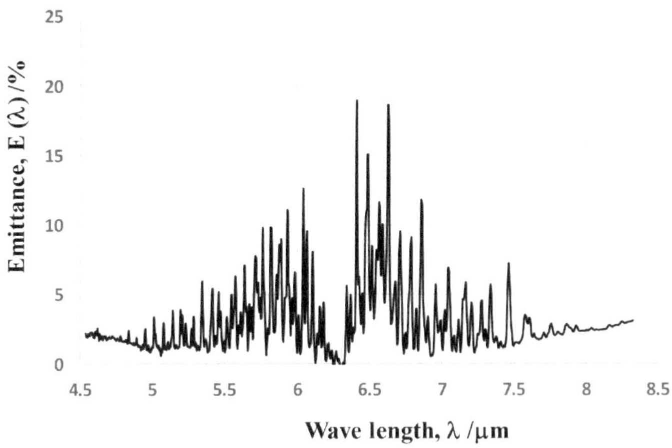

Figure 10.
The measured IR-emittance by the FT-IR for the LL-type specimen [22, 24, 25].

The IR-emittance E (λ) is characterized from the measured transmittance T (Λ) by FT-IR in the function of the wavenumber, Λ (cm-^{1}). Since λ (μm) = 10^4/Λ and E (λ) = 1.0 – T (λ), E (λ) represents the IR-emission from each type of acicular micro-textured specimens. No significant peaks were analyzed for λ < 4.5 μm and λ > 8.3 μm; T ~ 100% or E ~ 0% for λ < 4.5 μm and λ > 8.3 μm. In correspondence to the bimodal distribution of P (H) in **Figure 8a**, this E (λ) has bimodal profiles, the median of which is located at λ = 5.9 μm and λ = 6.7 μm, respectively. The standard deviation, Δλ, for two profiles is nearly the same as Δλ = 0.5 μm.

As discussed in [22], the measured spectra on the IR-emission at the five positions on the LL-specimen were compared among them. Although the measured intensities deviate from each other, the measure spectra were equivalent to E (λ) in **Figure 10**. This implies that the measured E (λ) by FT-IR is never dependent on the positions in measurement but intrinsic to the LL-type acicular microtextures in **Figure 3a** with the micro-cone unit cell size distribution in **Figure 8**.

Other three specimens with the same size as the LL-specimen were prepared for FT-IR measurement in the similar manner. The measurement zone was fixed to the center point of every specimen in practice. The wavelength range was only varied to indicate the difference in measured E (λ) among three specimens.

In both the LL- and M-type specimens, the IR-emission peaks are detected in their spectra as shown in **Figures 10** and **11a**. The wavelength of this peak shifts to the shorter wavelength, and the peak itself becomes too broad to detect the median of peak directly from the measured spectrum for the S-type specimen in **Figure 11b**. No IR-emission peaks are detected in the measured spectra for the SS-type specimen in **Figure 11c**. As seen in the SEM images on three specimens in **Figure 3**, the shorter micro-cone unit cells align with more regular and dense manner so that less free volume is present among the micro-cone unit cells. **Figure 11** implies that various micro−/nano-texturing parameters of micro-cone unit cells have influence on the IR-emission from the acicular micro-textured film, to be discussed in later.

The average wavelength of emission spectra, λ_{ave}, and the standard deviation Δλ in each E (λ) are estimated from **Figure 11** and listed in **Table 2**.

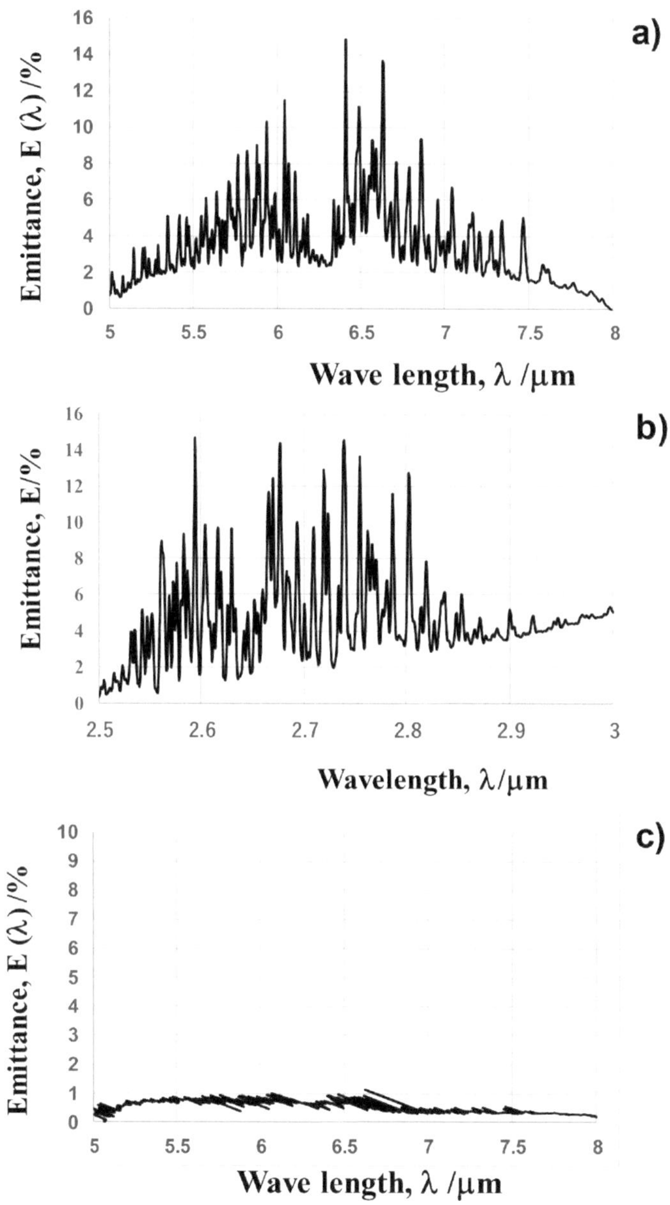

Figure 11.
The measured IR-emittance by the FT-IR for M-, S-, and SS-type specimens [25]. (a) E (λ) for M-type specimen, (b) E (λ) for S-type specimen, and (c) E (λ) for SS-type specimen.

Using the data in **Tables 1** and **2**, the correlation between $\{H_{ave}, \Delta H\}$ and $\{\lambda_{ave}, \Delta\lambda\}$ is investigated in **Figure 12**.

Within the measured standard deviation $\{\Delta H, \Delta\lambda\}$, the resonance relation for the IR-emission by the acicular microtextured device abides by $\lambda_{ave} = 2 \times H_{ave}$ when the emission wavelength is larger than 1 μm. In the IR-emission with its wavelength shorter than 1 μm, more precise measurement is necessary to investigate the effect of acicular microtexture size and topology on the IR-emission.

Acicular microtextured specimen	Average wavelength, λ_{ave}, in IR-emission (μm)	Standard deviation, $\Delta\lambda$, in IR-emission (μm)
LL-type specimen	6.7	0.25
LL-type specimen	5.9	0.25
M-type specimen	6.5	0.4
M-type specimen	5.7	0.4
S-type specimen	2.7	0.7
SS-type specimen	—	—

Table 2.
The average wavelength of emission spectra, λ_{ave}, and the standard deviation, $\Delta\lambda$ in the measured IR-emission spectra, E (λ), for four specimens.

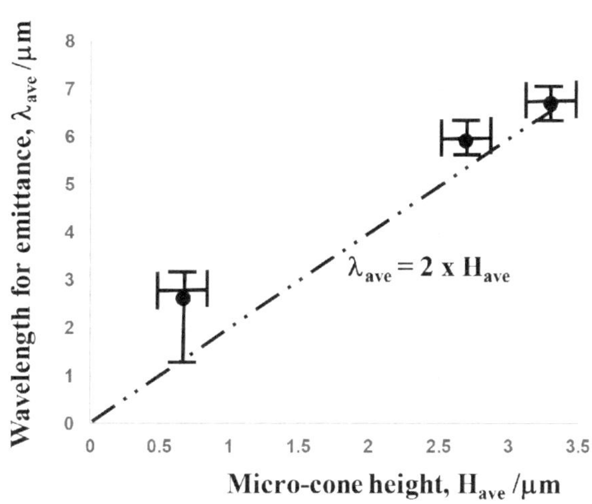

Figure 12.
Correlation of {λ_{ave}, $\Delta\lambda$} in the IR-emission and {H_{ave}, ΔH} in the population of micro-cone unit cell heights to deduce the resonance relation between the resonant IR-waves and the micro-cone unit cells.

3.3 Thermal transportation via heat radiation

The experimental setup in **Figure 6** was first utilized to describe the temperature difference between the heat source and the top surface of acicular micro-textured device. In the following experiment, the former temperature is controlled to be constant by 373 K, while the latter one is determined by the heat flux (q) balance between the q_{inout} from the heat source and q_{output} to the vacuum by the IR-emission. **Figure 13** depicts the stationary temperature distribution, measured by the thermography with its focus onto the LL-specimen surface.

The measured temperature on the acicular micro-textured LL-type specimen decreases down to 323 K, while the heat source surface temperature is constant by 373 K. Due to the IR-emission from the LL-type specimen, the IR-emitter surface temperature is reduced by ΔT = 50 K from the heat source temperature. In the latter, the IR-emission model is discussed with reference to [21, 22, 30]. The physical model for IR-emission from the acicular micro-textured copper plate is also considered for further understanding on the low-temperature heat radiation.

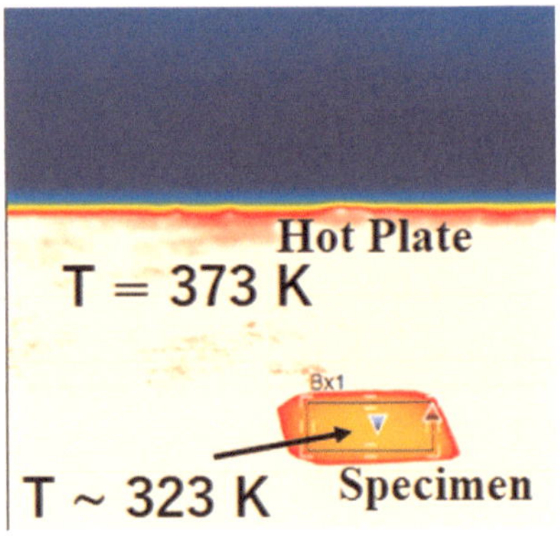

Figure 13.
Measurement of the stationary temperature distribution on the heat source (or hot plate) with the micro-textured LL-type specimen. The thermography was focused onto the surface of specimen.

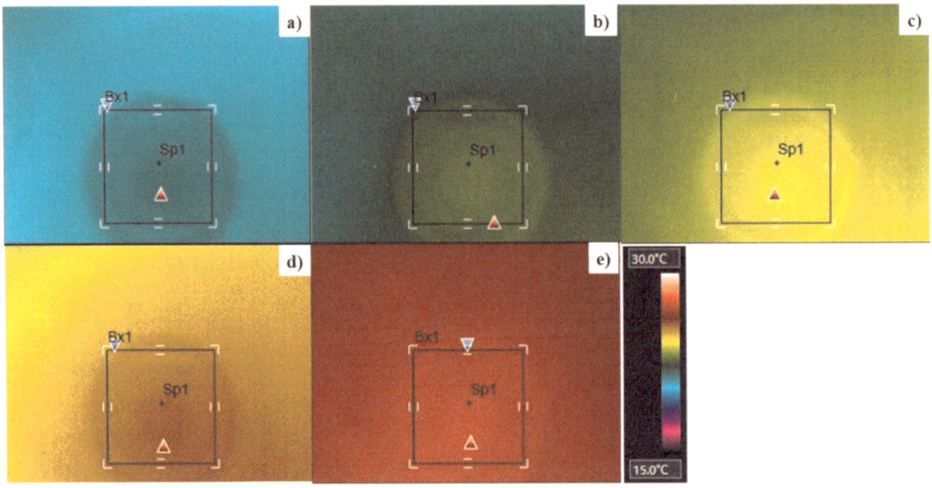

Figure 14.
Thermal transient on the top surface of BC-PC plate located away from the LL-type IR-emitter by 150 mm. (a) τ = 0 a, (b) τ = 60 s, (c) τ = 120 s, (d) τ = 360 s, and (e) τ = 600 s.

The heat radiation experiment is performed, using the experimental setup in **Figure 6**. Before evacuation, the BC-PC plate with the size of 180 mm x 180 mm x 2 mmt was placed with the distance from the LL-type micro-textured IR-emitter by 150 mm. The thermography was focused onto the top surface of BC-PC plate. **Figure 14** shows the thermal transient from the onset of measurement at τ = 0 s to the end of measurement at τ = 600 s. In experiment, the measurement starts when the hot plate temperature reaches to 293 K after switching it on; as shown in **Figure 14a**, the surface temperature starts to rise at the center zone of BC-PC plate. During 600 s, the temperature rise reaches 10 K on the top surface of objective body even when the heat source temperature is constant by 323 K as shown in **Figure 14b–e**.

Next, this LL-type specimen is exchanged with the S-type specimen to describe the heat radiation capacity change by reducing the micro-cone heights as shown in **Figures 3** and **12**.

Comparing **Figure 15** for the S-type IR-emitter with **Figure 14** for the LL-type IR-emitter, the thermal transient process is retarded and the final temperature at the

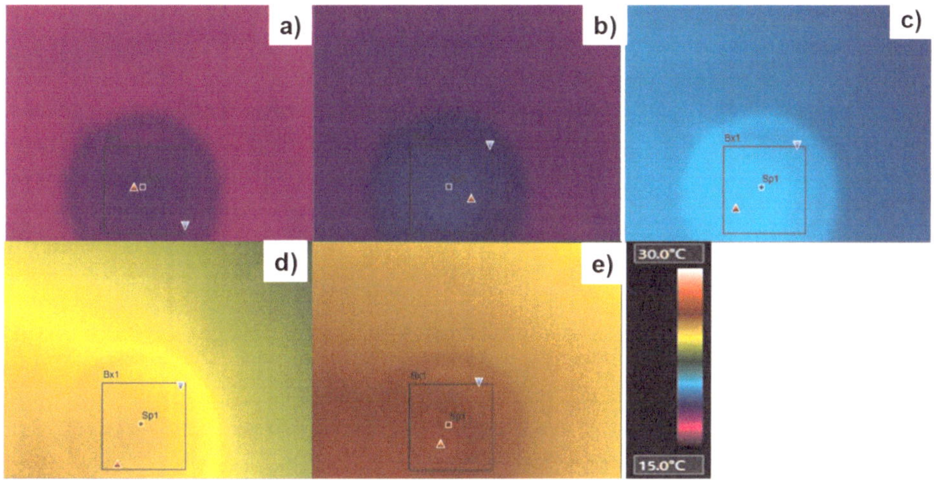

Figure 15.
Thermal transient on the top surface of BC-PC plate located away from the S-type IR-emitter by 150 mm. (a) τ = 0 a, (b) τ = 60 s, (c) τ = 120 s, (d) τ = 360 s, and (e) τ = 600 s.

hottest spot is reduced by 2 to 3 K. This implies that heat radiation capacity is directly affected by the IR-emission behavior with the micro-cone unit cell size dependency. More heat radiation experiments with higher micro-cone unit cells are necessary to demonstrate the possibility to improve the heat radiation capacity.

3.4 Cooling capacity of convection heat transfer under the forced cooling air

The heat convection under the forced cooling air is performed by using the experimental setup in **Figure 7**. The nozzle was fixed to blow the air in the lateral direction of the micro-textured and bare copper specimens. This air flow was switched on to cool down the specimen surface temperature and switched off to measure the recovery of surface temperature on the specimen. A simple theoretical model is used to describe these cooling and recovery processes of specimen surface temperature.

In this cooling mode, the surface temperature (T) reduces with time from the initial temperature (T_i) down to the lowest temperature (T_L), exponentially. This cooling behavior is modeled by.

$$(T-T_i)/(T_L-T_i) = 1-\exp.(-\kappa_{cooling} \text{ x } \tau), \tag{1}$$

where τ is the transient time from $T = T_i$ to $T = T_L$, and $\kappa_{cooling}$ represents the cooling capacity of specimen under the forced air-cooling condition. High $\kappa_{cooling}$ denotes for high cooling rate. This Eq (1) is utilized to estimate this $\kappa_{cooling}$ by parameter fitting by linear-logarithmically plotting the measured data in experiment. On the other hand, when the forced air blow is stopped, the surface temperature starts to recover from T_L to T_i. This recovery process is also modeled by.

$$(T-T_i)/(T_L-T_i) = \exp.(-\kappa_{recovery} \text{ x } \tau), \tag{2}$$

where $\kappa_{recovery}$ denotes for the recovery constant. In a similar manner to deduce $k_{cooling}$, this $k_{recovery}$ was estimated from the measured data.

Figure 16 summarizes the cooling and recovery responses in the convection heat transfer on the bare copper plate when switching on or off the air blow, respectively. Each experiment was performed three times to deduce the average response with the deviation of measured temperature less than 0.1 K. In a similar manner, four specimens were employed to measure the cooling and recovery responses and to compare the cooling and recovery capacities among four specimens with reference to the bare copper specimen.

Figure 17 depicts the cooling and recovery transients of acicular micro-textured LL-type specimen. Comparing **Figures 16** and **17**, the LL-type textured surface is rapidly cooled down to T_L < 313 K (or 40°C) and rapidly recovered to T_i. This reveals that $\kappa_{cooling}$ and $\kappa_{recovery}$ of textured copper plate in Eqs. (1) and (2) are more than κ's of bare copper plate.

Using these cooling and recovery capacities, the heat convection transfer process is compared among four micro-textured copper plates with reference to the bare copper plate. **Figure 18** shows the variation of cooling capacity or $\kappa_{cooling}$ of four textured specimens with increasing the micro-cone unit cell height. Each $\kappa_{cooling}$ was measured from the monitored temperature transient data after Eq. (1). When using the SS- and S-specimens, this cooking capacity is enhanced by their dense alignment of shorter micro-cone unit cells. This enhancement is attributed to the enlargement of heat transfer area with reduction of micro-cone size. On the other hand, this capacity also

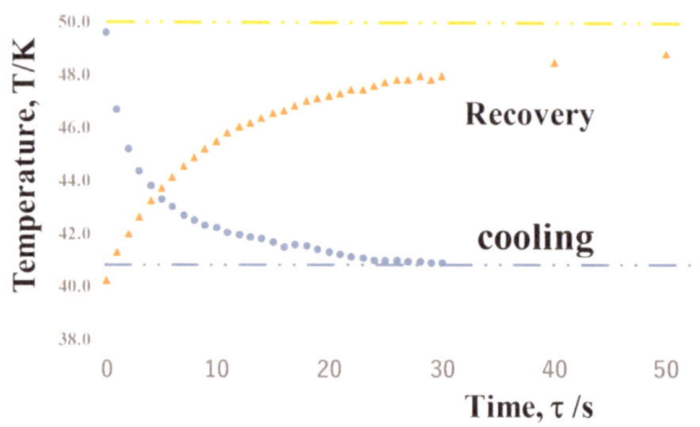

Figure 16.
Cooling and recovery responses in the convection heat transfer of bare copper plate when switching on or off the forced air blow, alternatively.

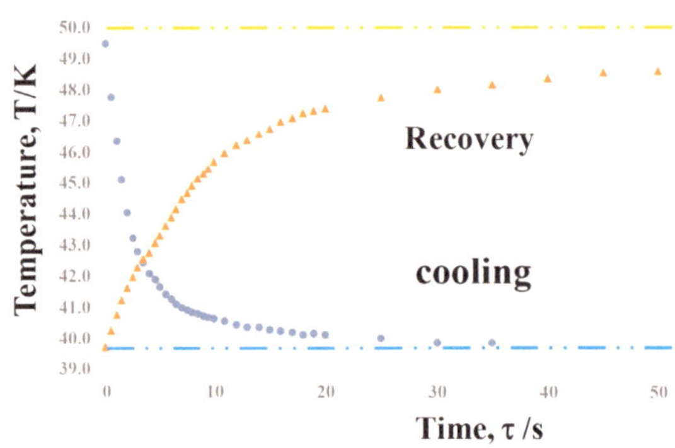

Figure 17.
Cooling and recovery responses in the convection heat transfer of the micro-textured LL-type specimen when switching on or off the forced air blow, alternatively.

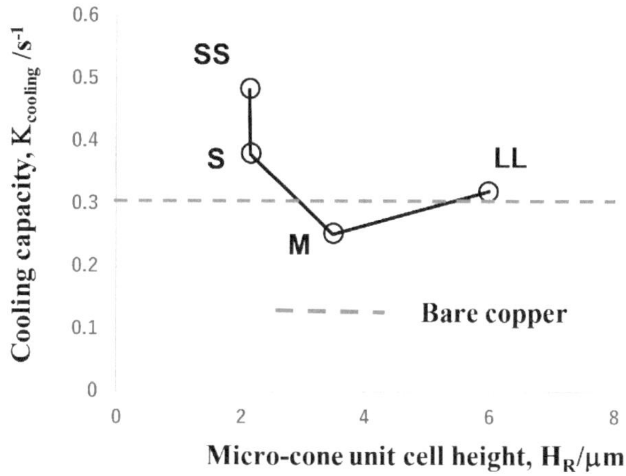

Figure 18.
Comparison of the cooling capacity among four specimens with reference to the bare copper specimen. The representative micro-cone unit cell height, H_R, is equal to H_{ave} for specimens with mono-modal distribution of P (H) and defined by $H_R = (H_{ave}^1 + H_{ave}^2)/2$ for specimens with bimodal distribution of P (H).

increases with increasing the micro-cone unit cell heights. To be remembered, the free volume increases with the micro-texture heights; the micro-cone unit cell density decreases when using the M- and LL-specimens. This implies that the convection heat transfer is enhanced by the forced cooling flow through the microtextures. The convection heat transfer mechanism is expected to change with the acicular microtextures with optimum alignment of micro-cone unit cells and FVEs.

4. Discussion

The acicular micro-textured thin structure with semi-regular alignment of micro-cone unit cells has an IR-emission capability from low-temperature heat source. This IR-emission is driven by the electromagnetic wave resonance with the free volume elements (FVEs) among the acicular microtextures. Since each FVE is surrounded by the micro-cone unit cells, this FVE is wrapped by the metallic walls with high intensity heat flux, as illustrated in **Figure 19a**. Let us estimate this heat flux (q) for IR-emission from the LL-specimen on the heat source. This q is given by $q = -\kappa_a \times (T_S - T_B)/h$, where κ_a is the thermal conductivity of microtextured film, h is the film thickness, and T_S and T_B are the surface and bottom temperatures of micro-textured structure, respectively.

Since the micro-textured structure consists of the Fe-Ni alloy, this κ_a is estimated to be 80 W / (m · K). In case of the LL-type specimen, h = 4 μm. In the experimental setup in **Figures 6** and **13**, T_B is equal to the hot plate temperature, 373 K, and T_S is the measured temperature, 323 K. Then, each heat flux surrounding the FVE reaches to 1 GW / m^2 in the physical model in **Figure 19b**. This power dissipates to induce the electromagnetic waves with both nodes at the bottom and top surfaces of free volume elements, respectively. The resonance relationship by $\lambda_R = 2 \times H_R$ in **Figure 12** proves that the electromagnetic waves with the wavelength of λ_R is induced at each FVE among the micro-cone unit cells in the micro-textured film.

In [21, 30], the micro-cavity aluminum sheet was used as an IR-emitter in the heat radiation experiment. The micro-cavity unit cell was sized by 3.5 μm x 3.5 μm x 5 μm.

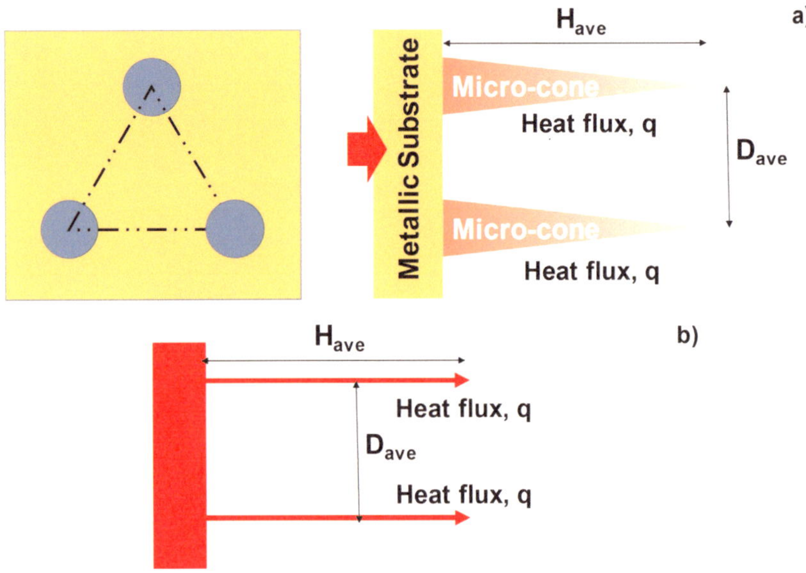

Figure 19.
Physical model on the infrared electromagnetic wave resonance in the inside of the free volume, surrounded by the micro-cone unit cells in the micro-textured structure. (a) A schematic view of acicular microtextured IR-emitter, and (b) a physical model of IR-emission.

T_S and T_B were also measured to be (T_S-T_B) = 10 K. The IR-emission from this micro-cavity aluminum sheet was detected with reference to the bare aluminum sheet; however, the intensity of IR-emission was still limited in practice. This might be attributed to low resonation of micro-cavity alignment to emit the IR-waves.

Figure 20 illustrates the role of free volume in the IR-emission. When the FVE with the finite volume is present among the micro-cone unit cells, this FVE works as a resonator to emit the IR-waves with the defined wavelength, λ_R, by the resonator height. Notably, the emitted IR-waves simultaneously direct in the vertical direction to the copper substrate in this device. The regularity of micro-textured micro-cone unit cells has much influence on this directivity of IR-waves.

Before further discussion, let us consider the role of free volume element in organic and inorganic materials. As seen in the literature, the free volume element (FVE) works as a key item to determine the functional properties of materials. The free volume fraction plays a key role in the network structure of oxide glasses [31]. In particular, this fraction has a close relationship to the fictitious temperature in the structural relaxation, or to various mechanical and functional properties of oxide glass

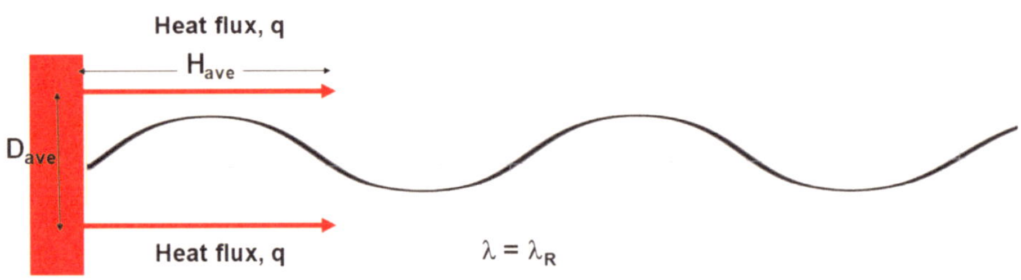

Figure 20.
IR-wave emission from each FVE to the vacuum through the resonator voxel. The voxel height defines the emitted IR wavelength.

systems [32]. In the polymer materials such as low-density PMMA (Poly-Methyl--Meta-Acrylate), the nano-scaled FVE induces the ultrafast structural fluctuation in the inside [33]. These findings on the significant role of FVE in various material systems reveal that FVE with a finite volume in the acicular microtextures is responsible for IR-emission through the resonance to drive the electromagnetic waves by the dissipated thermal power to FVE.

Let us consider the micro-cone unit cell size effect on this IR-emission. The resonance relationship in **Figure 12** and the measured IR-emissivity in **Figures 10** and **11** demonstrate that IR-emission is induced for LL-, M-, and S-specimens but that no IR-emission takes place for SS-specimen. SEM images on the microstructures in **Figure 3** show that a free volume is present in the micro-textured film of LL- and M-type specimens but that the micro-cone unit cells align by themselves too densely to induce the electromagnetic waves. This size effect on the IR-emission is physically described by using the FVE model. As illustrated in **Figure 21**, the integrated area ratio, A, of IR-emission spectra per the whole integrated intensity in the all wavelength range decreases with the free volume fraction, f. When using the LL-, M-, and S-specimens, the IR-waves are emitted since f is more than the critical free volume fraction (f_c). However, no IR-emission is detected in **Figure 11c** in the case of the SS-specimen since $f < fc$.

Various studies were reported in the literature on the microstructural design toward high optical antireflection films [34]. Optical ray simulation was utilized to describe the relationship between the optical reflectivity and the micro−/nano-textured surface. COMSOL software was also utilized to design the anti-reflecting microstructures for IR-applications [35]. The effect of microstructural unit cell shape on the antireflection and its broad band was discussed through the simulation. These studies suggest that the physical model in **Figure 19** is installed into the multi-physics model with consideration to the resonance between the heat dissipation and the IR-emission.

The acicular microtextures control the heat convection mechanism. As stated before, the alignment of micro-cone unit cells or the related size effect to D_{ave} has direct influence on the cooling behavior. With decreasing D_{ave}, the unit cell density increases; the extended surface area enhances the convection heat transfer as depicted in **Figure 22a**. On the other hand, the coolant flow has a structure where the viscous layer at the vicinity of walls has little influence from the main flow. After [13, 14],

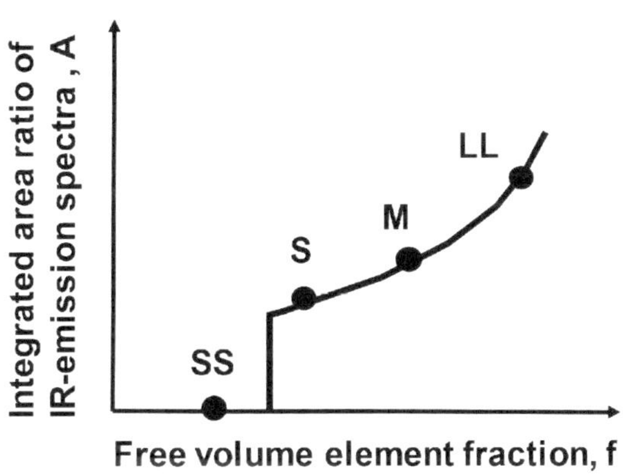

Figure 21.
Fee volume element model in the acicular microtextured with different micro-cone unit cell sizes.

when the acicular microtextures are within this viscous layer or the unit cell height is smaller than the viscous layer thickness (y+), the heat transfer through the microtextures has nothing to do with the main flow of coolant. As depicted in **Figure 22b**, when $H_{ave} > y+$, the main flow pattern is affected by the acicular microtextures so that the convection heat transfer changes by itself at the presence of microtextures. Three dimensional flow dynamic simulation [36] might be useful to describe this interaction between the coolant flow and the micro-cone unit cells and to theoretically understand the microtexturing effect on the convection heat transfer.

In addition to the microtexture size effect on the convection heat transfer, the topological design is also needed to make full use of acicular microtextures for heat transfer control. In the case of the convex-type acicular microtextures in **Figure 23a**, H_{ave}/B_{ave} and D_{ave}/B_{ave} ratios in {H, B, D}-design influence on the heat transfer from the conductively heated micro-cone surfaces to the forced coolant flow. As partially stated in [13, 14], the nucleation site selection of vapors on the micro-cone unit cell surface influences on the bubble density during the boiling heat transfer. This relationship between the acicular microtextures and the vapor nucleation changes by itself by topological change of the convex textures to the concave ones in **Figure 23b**. After the classical boiling heat transfer treatise [9], the vapor was assumed to nucleate itself at the bottom of concave acicular microtextures and to grow in them during the nucleation step in the boiling curve. This suggests the transition of nucleation-to-growth steps of vapors is significantly affected by the concave microtexture sizes.

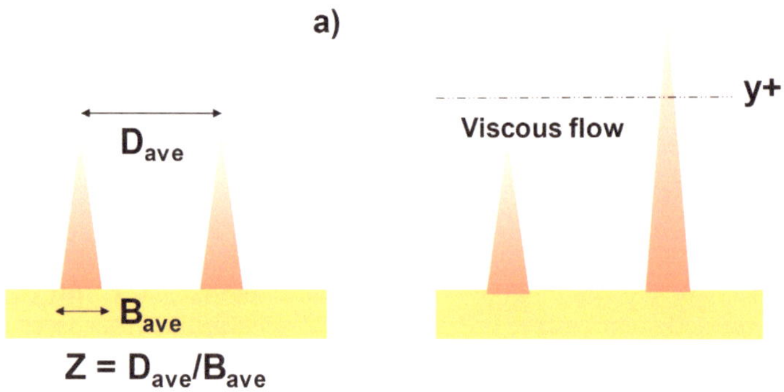

Figure 22.
Two types of the microtexturing size effect on the convection heat transfer under the forced coolant flow. (a) Dimensional relationship among {H, B, D} in the acicular microtextures, and (b) relationship between the micro-cone height and the viscous flow layer thickness, y+.

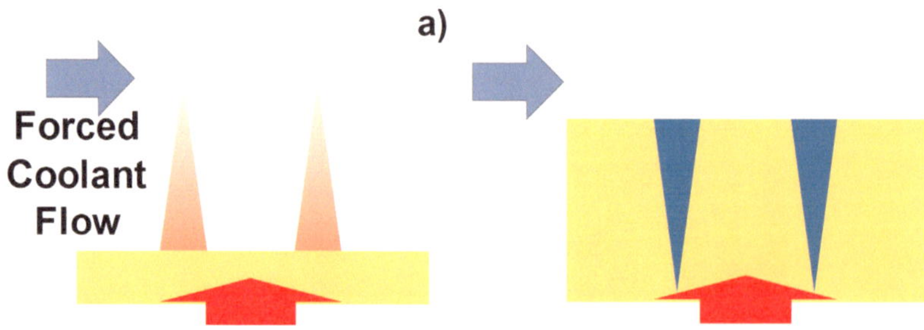

Figure 23.
Comparison of the convex and concave acicular microtextures at the convection heat transfer. (a) Convex acicular microtextures, and (b) concave acicular microtextures.

In addition to the acicular microtexture size effect on the IR-emission, let us consider how the micro-cone unit cell regularity influences on this IR-emission. As had been studied in [19, 20], the absorbance or emittance of IR-waves by the nano-structured and meta-material devices was preserved to be nearly 100% in the tailored wavelength range from 1 μm to 5 μm. As shown in **Figure 10**, the maximum emittance is limited by 20 to 30%. This inefficient IR-emission by semi-regularly aligned micro-cone unit cells is attributed to their dispersive distribution with the topological irregularity of FVE. In particular, the geometric and topological distortion of FVEs might significantly reduce the IR-emittance.

The geometric and topological regularity of micro-cones is much improved by the nucleation site control as depicted in **Figure 2**. Their optimum size and shape is investigated to minimize the distortion of FVEs after acicular microtexturing.

Let us summarize the new applications of acicular micro-textured film for thermal transportation and heat transfer.

In the present device, the IR-wave is emitted from each FVE, surrounded by the high heat fluxes. When every micro-cone unit cell is formed onto the curved substrate surface with designated curvature, these emitted IR-waves are focused onto the objective body surface to heat it up by this radiation process, as illustrated in **Figure 24a**. As stated in [35, 37], the acicular micro-textured film with more regular alignment of micro-cone unit cells works as an antireflective device to control the refractive index by its size effect. In **Figure 24b**, the input IR-wave E_1 is transformed to an output IR-wave E_2 by the optical control through the film.

The heat pipe was invented to make long-distance heat transfer from the heat-in section *via* evaporation to the heat-out section *via* condensation [38]. Its inner surface is made of the capillary wick to house the condensate film and the vapor. Preserving nearly the same structure as heat pipe, the vapor chamber is designed as a thin, flat structure unit to make two-phase heat transfer [39]. In its conventional design, the wick is made of the sintered copper layer and copper fins. In both devices, the capillary wick works as an essential structure for evaporation, condensation, and mass flow.

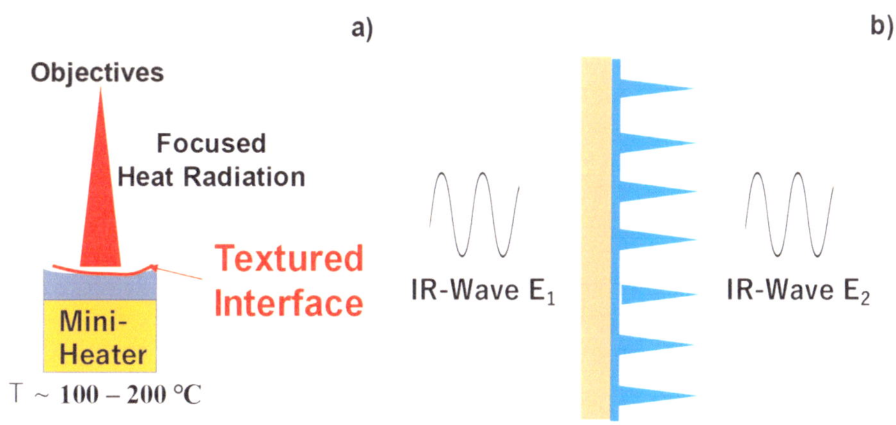

Figure 24.
Controllability of radiation heat transfer process and IR-wave transformation. (a) Focusing of emitted IR-waves from the curved microtextured film to the objective body, and (b) antireflection with the tailored refractive index for transformation of IR-waves.

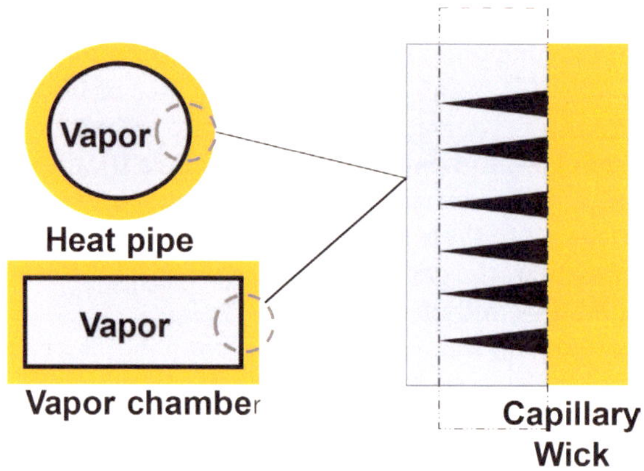

Figure 25.
Controllability of convection heat transfer process in the inside of heat pipe and vapor chamber units.

The inner surfaces of heat pipes and vapor chambers are easily wet-plated by using the present electrochemical procedure, as depicted in **Figure 25**. The micro-cone unit cell sizes are also controllable just in the similar manner as stated before. These Fe-Ni alloy acicular microtextures have sufficient strength and toughness as a structural member. They provide the dense nucleation sites for efficient evaporation and condensation at lower superheat. The flowability of condensate liquids is easily controlled by the micro-cone unit cell sizes.

5. Conclusion

Infrared (IR) optics to terahertz electromagnetics is a new technological field to cultivate the innovative ways onto the basic science and to harvest the valuable applications in heat transfer. The micro–/nano-texturing provides a linkage between this basic understanding on the IR-emission and the heat transportation through radiation. The free volume element (FVE) among the micro-cone unit cells plays a role of IR-emission with the selective wavelength through the resonance of IR-electromagnetic waves to the micro-cone unit cells or the FVE. Since each FVE is surrounded by micro-cone cells with high heat flux, the IR-waves are emitted to have two nodes at the bottom and top of FVE. The wavelength (l) in this IR-emission is determined to be equal to the doubled FVE-length or the doubled micro-cone heights.

At the presence of resonant free volume element in the acicular micro-textured film, the wavelength of emitted IR electromagnetic waves is controllable from near-IR to far-IR range by varying the micro-cone height. With decreasing the height, the micro-cone unit cells are densely formed on the substrate and the free volume reduces so that IR-emittance significantly decreases and disappears. This size effect of acicular microtextures with semi-regular alignment suggests that more regular alignment is needed to preserve high IR-emittance in the wider selective range of resonant wavelengths.

This acicular microtexturing provides a new way to describe the convection heat transfer process within the viscous layer of coolant or with interaction to the main coolant flow. Various size effects on the single-phase or the two-phase heat transfer between the heating solid and the coolant are experimentally analyzed by using the present device.

In the heat radiation by this acicular micro-textured device, the object located away from the IR-emitter is efficiently heated up till the thermal balance around the objective body. In the heat convection, higher cooling rate is attained by using this acicular micro-textured device. These micotextures are formed on the curved surfaces even with large to be applied to radiation- or convection-based heat transfer of structural members and parts with complex shape and geometry. These are easily mounted and installed as a capillary wick into various heat sinks, heat pipes, and vapor chambers. Phase transformation of cooling media through evaporation and condensation processes is much enhanced together with the control of condensate liquids.

Acknowledgements

The authors would like to express their gratitude to Prof. Ono N. (Shibaura Institute of Technology) and Kurozumi S-I. (Nano-Film Coat, llc.) for their advice and help in experiments.

Conflict of interest

The authors declare no conflict of interest.

Nomenclature

A	integrated area ratio of IR-emission spectra in whole spectrum
B	root diameter of micro-cone unit cells
B_{ave}	median of measured B
ΔB	standard deviation of measured B
D	mutual distance between adjacent micro-cone unit cells
D_{ave}	median of measured D
ΔD	standard deviation of D
E (λ)	emittance in the wavelength, λ
H	height of micro-cone unit cell
H_{ave}	median of measured H
ΔH	standard deviation of H
P	population of micro-cone unit cell size parameters {H, D, B}
T (Λ)	transmittance in the wave number, Λ
T (λ)	transmittance in the wavelength, λ
q	heat flux
Λ	wave number in cm^{-1}
l	wavelength in μm
lave	median of measured wavelengths
Δλ	standard deviation of measured wavelengths

Author details

Tatsuhiko Aizawa[1*], Hiroki Nakata[2] and Takeshi Nasu[2]

1 Surface Engineering Design Laboratory, Shibaura Institute of Technology, Tokyo, Japan

2 Ebina Denka Kogyo, Co., Ltd., Tokyo, Japan

*Address all correspondence to: taizawa@sic.shibaura-it.ac.jp

References

[1] Ganji DD, Sabzehmeidani Y, Sedghiamiri A. Nonlinear Systems in Heat Transfer. London, UK: Elsevier; 2018

[2] Frankerwicz M, Golda A, Kos A. Investigation of heat transfer in integrated circuits. Metrology Measurement System. 2014;**11**(1): 111-120

[3] Gonga C, Shen J, Yu T, Wang K, Tu Z. A novel radiator structure for enhanced heat transfer used in PEM fuel cell vehicle. International Journal of Heat and Mass Transfer. 2020;**157**:119926

[4] Liang J, Wu J, Guo J, Li H, Zhou X, Liang S, et al. Radiative cooling for passive thermal management towards sustainable carbon neutrality. National Science Review. 2022;**10**(1):208

[5] Saito Y, Aizawa T, Wasa K, Nogami Y. Leak-proof packaging for GaN chip with controlled thermal spreading and transients. In: Proceedings of the 2018 IEEE BiCMOS and Compound Semiconductor Integrated Circuits and Technology Symposium (BCICTS), 15–17 October. San Diego, CA, USA; 2018. pp. 243-246

[6] Aizawa T, Wasa K, Nogami Y. Plasma oxidation printing into DLC and graphite for surface functionalization. Journal of Carbon Research. 2019;**5**(11):1-10

[7] Aizawa T, Nakata H, Nasu T, Nogami Y. Micro-textured graphitic substrate—copper packaging for robustness. In: Proceedings of the 5th WCMNM Conference, September 22, 2022. Leuven, Belgium; 2022. pp. 291-295

[8] Aizawa T, Nakata J, Nasu T, Nogami Y. Robust packaging of vertically aligned graphite substrate by copper micro-rib structuring. Journal of Carbon Research. 2022;**8**(70):1-13

[9] Kattoh Y. Fundamentals in Heat Transfer. Tokyo, Japan: Yokendo; 1964

[10] Ahn HS, Lee C, Kim H, Jo H, Kang S-H, Kim J, et al. Pool boiling CHF enhancement by micro/nanoscale modification of zircaloy-4 surface. Nuclear Engineering Design. 2010;**240**: 3350-3360

[11] Jung SM, Preston DJ, Jung HY, Deng Z, Wang EN, Kong J. Porous Cu nanowire aerosponges from one-step assembly and their applications in heat dissipation. Advanced Materials. 2016; **28**:1413-1419

[12] Aizawa T, Wasa K, Tamagaki H. A DLC-punch array to fabricate the micro-textured aluminum sheet for boiling heat transfer control. Journal of Micromachines. 2018;**9**(147):1-10

[13] Aizawa T, Ono N. Boiling heat transfer control by micro−/nano-texturing of metallic heat-spreading devices. In: Proc. 3rd WCMNM2021, September 21, 2021. Mumbai, India; 2021. pp. 1-10

[14] Aizawa T, Ono N, Nakata H. Boiling heat transfer on the micro-textured interfaces. Heat Transfer. 2022;**2022**: 61-86

[15] Khalil M, Hassan-Ali MI, Khan KA, Al-Rub RA. Forced convection heat transfer in heat sinks with topologies based on triply periodic minima surfaces. Case Studies in Thermal Engineering. 2022;**38**:102313

[16] Gilmore N, Timchenko V, Menictas C. Manifold microchannel heat

sink topology optimization. International Journal of Heat and Mass Transfer. 2021; **170**:121025

[17] Aizawa T, Nakata H, Nasu T. Manufacturing and characterization of acicular Fe-Ni micro-textured heat-transferring sheets. In: Proc. WCMNM2022 , September 21, 2022. Leuven, Belgium; 2022. pp. 277-281

[18] Bergman TL. Active daytime radiative cooling using spectrally selective surfaces for air conditioning and refrigeration. Solar Energy. 2018; **174**:16-23

[19] Li W, Fan S. Nano-photonic control of thermal radiation for energy applications. Optics Express. 2018;**26**: 15995-16021

[20] Tian Y, Ghanekar A, Ricci M, Hyde M, Gregory O, Zheng Y. A review of tunable wavelength selectivity of metamaterials in near-field and far-field radiative thermal transport. Materials. 2018;**11**:862

[21] Kohiyama A, Shimizu M, Yugami H. Unidirectional radiative heat transfer with a spectrally selective planar absorber/emitter for high-efficiency solar thermos-photovoltaic systems. Applied Physics Express. 2016;**9**: 112302

[22] Aizawa T, Nakata H, Nasu T. Fabrication and characterization of acicular micro-textured copper sheet device for low-temperature heat radiation. Journal of Micromachines. 2023;**14**(507):1-12

[23] Aizawa T, Nakata H, Nasu T. Acicular microtexturing onto copper sheet for heat radiation. In: Proc. 12th AWMFT Conference, May 17, 2023. Gangneung, Korea; 2023. p. 67

[24] Aizawa T, Nakata H, Nasu T, Mitohka Y. Acicular microtextured sheet device for heat radiation with controlled unit-cell size and regularity. In: Proceedings of the 5th WCMNM Conference, September 20, 2023. Evanston, USA; 2023. pp. 5-99 (in press)

[25] Aizawa T, Nakata H, Nasu T. Infra-red emission from micro-cone textured metallic sheet with semi-regular alignment. Journal of Photonics. 2023 (in press)

[26] Aizawa T, Inohara T. Pico- and femtosecond laser micromachining for surface texturing. In: Micromachining. London, UK, London, UK: IntechOpen; 2019. pp. 1-24

[27] Aizawa T, Inohara T, Wasa K. Fabrication of superhydrophobic stainless steel nozzles by femtosecond laser micro−/nano-texturing. International Journal of Automation Technology. 2020;**14**(2):159-166

[28] Oho E. Digital image processing technology for scanning electron microscopy. Advances in Imaging and Electron Physics. 1998;**105**(77-112): 113-140

[29] Preparata FP, Shamos MI. Computational Geometry. New York, USA: Springer; 1985

[30] Okitsumo Co., Ltd. Development of Net Heatsink Type Thermal Emission Materials. Japan: METI; 2013. pp. 1-54

[31] Suzuki S, Takahashi M, Hikich Y. Free volume of network-forming oxide systems P_2O_5 – (GeO_2, TeO_2, Sb_2O_3, V_2O_5). Journal of the American Ceramic Society. 1987;**70**(9):C213

[32] Nishida T, Kubuki S, Oka N. Local structure, glass transition, structural relaxation, and crystallization of

functional oxide gasses investigated by Moessbauer spectroscopy and DTA. Journal of Material Science. 2021;**32**: 23655-23689

[33] Hoffman DJ, Fica-Contreras SM, Fayer MD. Amorphous polymer dynamic and free volume element size distributions from ultrafast IR spectroscopy. PNAS. 2020;**117**(25): 13949-13958

[34] Chen T, Wang W, Pan A, Hu L, Mei X. Characterization of anti-reflection structures fabricated via laser micro/nano-processing. Optical Materials. 2022;**131**:112686

[35] Pal U. Designing antireflecting microstructures for infrared applications. COMSOL Blog. Available from: https//:www.comsol.com/blogs/designing-antireflecting-microstructures-for-infrared-applications/ [Accessed May 23, 2023]

[36] Yamada T, Ono N. A study on micromixing utilizing Marangoni effect induced on gas-liquid free interfaces. Journal of Micro-/Nano-manufacturing. 2015;**3**:021003

[37] Lopez-Alonso JM, Mandviawala T, Alda J, Lail B, Boreman G. Infrared antenna metrology. Proceedings of SPIE. 2005;**5987**:59870L1-59870L11

[38] Bahman Z. Basic principles of heat pipes and history. In: Heat Pipe Design and Technology. Springer; 2016. pp. 1-41

[39] Peterson GP, Ma H. Chapter one – Advances in vapor chambers and phase change heat spreaders. Advances in Heat Transfer. 2021;**53**:1-96

Chapter 7

Analysis and Optimization of Heat Transport for the Purpose of Maximizing the Potential of Solar Ponds in Sustainable Energy Applications

Baljit Singh Bhathal Singh

Abstract

Modern solar ponds can be used for power generation, water heating, and even desalinization. Understanding the internal heat transport mechanisms is vital for maximizing their potential use. In-depth discussion of how to analyze heat transfer in solar ponds is provided in this chapter. Heat is transferred effectively in solar ponds by conduction, convection, and radiation. They do this by making salinity gradients, or layers, that absorb and store solar heat. Sunlight is absorbed by the upper layer, while the lower layers provide insulation. Researchers look into heat transfer in solar ponds using analytical, computational, and experimental approaches. Temperature distributions and heat transport rates are modelled mathematically using energy balance equations and fluid dynamics. Flow patterns and convective heat transfer are studied by CFD models. Understanding the efficiency of solar ponds is made easier by experimental observations of temperature profiles and heat flows. The importance of heat transfer analysis in determining the best values for design factors including pond depth, salinity gradient, and insulating materials is highlighted in this chapter. Sustainable potential of solar ponds in diverse energy applications can be unlocked by advancing our understanding of heat transport mechanisms and building accurate models.

Keywords: solar ponds, heat transfer analysis, thermal gradient, energy efficiency, sustainable energy applications

1. Introduction

It is possible to collect and store heat from the sun using natural solar thermal collectors such as solar ponds [1, 2]. Using temperature and density gradients, seawater is separated into three distinct layers in a large shallow pond. The densest, bottom layer of the pond soaks up the sun's rays and uses them to maintain a high temperature, which is subsequently transferred to the more vaporous upper layers through natural convection [3].

The idea of solar ponds was first proposed in the early 1950s [4] by French physicist Felix Trombe and Israeli scientist Israel Dostrovsky. But in 1960, American physicist John H. Reynolds constructed the first functional solar pond at Texas A&M University [2].

The results of a heat transfer analysis of a solar pond can be used to enhance the effectiveness of the installation. If engineers understand how heat travels through the pond, they can determine the most efficient way to pull it out and turn it into usable electricity. Problems like as heat loss, stratification, and stagnation can be discovered, and potential remedies proposed, using heat transfer analysis. Engineers can improve the efficiency and cost-effectiveness of solar energy systems by understanding more about heat transmission in solar ponds [1, 3].

2. Solar pond design and components

It has been shown that a solar pond has three distinct layers: the upper convective zone (UCZ), the non-convective zone (NCZ), and the lower convective zone (LCZ) [5] as shown in **Figure 1**. The uppermost layer consists of either freshwater or low-salt brine. This upper layer becomes hotter and less dense as a result of solar radiation. Naturally occurring convection currents arise as a result, removing the heat from the convective region [6]. The non-convective zone's water contains a high percentage of salt. It does not combine with the higher convective zone because of its high density, which is produced by the high concentration of salt. The storage space underneath benefits from this extra layer of protection from the elements. The

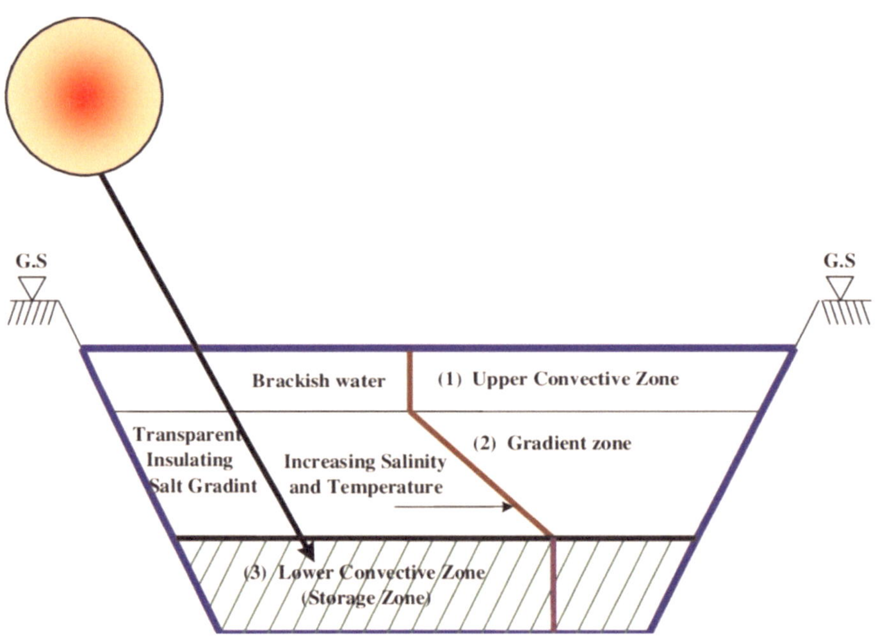

Figure 1.
Salinity gradient solar pond.

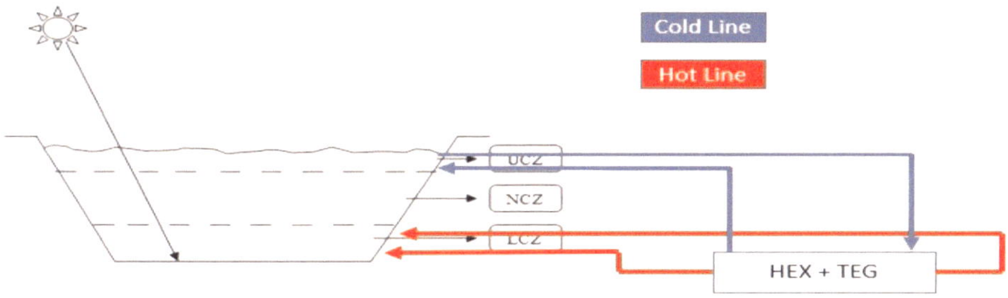

Figure 2.
Design of power generation from solar pond using thermoelectric generator.

pond's storage portion contains highly concentrated seawater. As shown in **Figure 2**, the heat from this storage zone can be extracted using heat exchanger and thermoelectric generators. The heat produced in the higher convective and non-convective zones is stored here. The thickness of the storage zone increases as the temperature of the salt rises due to the energy being stored [6, 7]. **Figure 3** shows the types of solar ponds relevant to heat transfer analysis.

Depending on whether a convective zone is present, solar ponds are either convective or non-convective. The well-mixed upper layer of convective solar ponds is heated by direct exposure to sunlight. The water is heated, rises to the surface, and then sinks again when it cools due to convection [7]. However, unlike convective solar ponds, which lose heat through convection, non-convective solar ponds feature a unique non-convective layer that functions as an insulator and keeps heat in the storage area. Because of the density gradient caused by the salt concentration, mixing between the higher convective layer and the non-convective layer is suppressed in the non-convective layer. Application and design factors should be considered while deciding between a convective and non-convective solar pond. For applications that need for high-temperature heat storage or when water is in short supply, non-convective ponds may be the way to go [8].

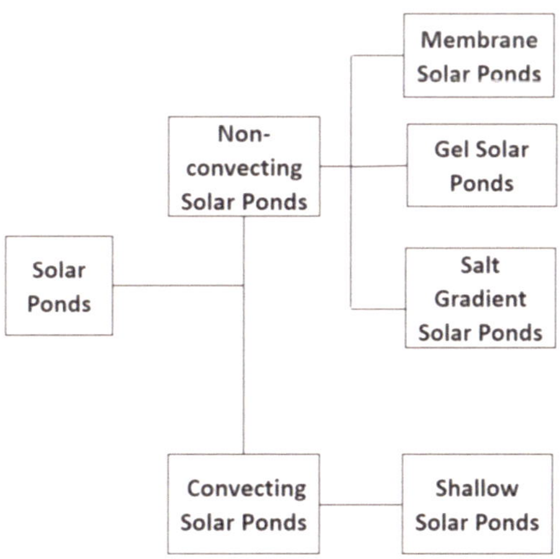

Figure 3.
Types of solar pond.

3. Factors affecting the performance of a solar pond relevant to heat transfer analysis

Both the depth and surface area of the pond influence the pace of convective heat transfer and the volume of water available for solar energy absorption [9]. Water temperature and salinity affect the rate of thermal energy storage and the density gradient [8]. The amount of solar energy absorbed by the pond is dependent on both the intensity and duration of solar radiation [10], while the rate of convective heat loss is regulated by air conditions. Overall, a good grasp of these factors is vital since they are critical for maximizing the design and operation of solar ponds for maximum solar energy utilization [11].

The design and operation of solar ponds are influenced by factors such as the pond liner material and heat extraction mechanism. Using high-quality insulation materials to line ponds, for instance, can reduce heat loss and improve thermal efficiency [12]. The efficiency with which thermal energy is converted to electrical energy, however, may depend on the heat extraction technique finally selected. Popular heat extraction techniques include heat pumps, absorption refrigeration systems, and heat exchangers [10]. If these factors can be optimized, solar ponds can be used to create successful and affordable solar energy systems.

4. Heat transfer in solar pond

The ability of solar ponds to capture and store thermal energy for long periods makes them a viable option for solar energy storage. It is crucial to consider the mechanics, governing equations, and influencing elements of heat transport in a solar pond while planning for and operating one [13]. The temperature, salinity, and velocity distributions in a solar pond are governed by a set of partial differential equations that represent heat transfer by conduction, convection, and radiation. The continuity, momentum, energy, and salt concentration equations are just some of the equations that can be solved numerically with finite element techniques [14]. Pond shape, salinity gradient, solar radiation intensity and duration, meteorological conditions, and heat exchanger design all have a role in the effectiveness of a solar pond. Heat transfer is also affected by things like the type of fill used, the depth of the water, and the movement of the water layers. To maximize solar pond performance and guarantee their efficacy in providing sustainable energy solutions [15], an understanding of these elements is crucial.

For a solar pond to function properly, heat transfer must be optimized. Heat can be transferred in a solar pond through conduction, convection, and radiation. These events can be described in terms of the partial differential equations that govern the distributions of temperature, salinity, and velocity in the pond [16]. The effectiveness of a solar pond can be affected by several factors, including the pond's design, the salinity gradient, the quantity of sunshine available, the weather, and the efficiency of the heat exchanger. You need to have a strong grasp on these factors [17] to get the most out of a solar pond. The continuity equation describes the mass conservation and velocity distribution inside the pond, whereas the momentum equation describes the flow of fluid and the forces acting against it [18]. The energy equation characterizes the temperature distribution and the heat transport within the pond. A salt concentration and its distribution in the pond are described by an Eq. [19]. A solar pond's equations need to be numerically solved in order to make reliable performance predictions. These issues have been addressed using a wide variety of numerical techniques [20], including the finite element method, the finite difference method, and the control volume method.

5. Mathematical modeling of heat transfer in a solar pond

The equation that may be used to calculate the heat extraction from the lower convective zone is shown below.

$$Q_{extraction} = m_{brine} \times c_p \times (T_{brine\ out} - T_{brine\ in}) \tag{1}$$

where,
m_{brine} = Mass flow rate of brine solution.
$c_{p,brine}$ = Specific heat of brine solution.
$T_{brine,out}$ = Temperature of the brine water as it leaves the heat exchanger.
$T_{brine,in}$ = The temperature at which brine water enters the heat exchanger.
The equation representing heat storage in the storage zone:

$$Q_{LCZ} = Q_I - Q_{HR} - Q_{hlg} - Q_{conv-loss} - Q_{rad-loss} - Q_{evap-loss} \tag{2}$$

where,

$$Q_{hlg} = k_G \times A \times (T_{LCZ} - T_G)/x_G \tag{3}$$

$$Q_{conv-loss} = h_c \times A \times (T_{UCZ} - T_A) \tag{4}$$

$$Q_{rad-loss} = \sigma \varepsilon_w \left\{ (T_{UCZ} + 273)^4 + \left(T_{sky} + 273\right)^4 \right. \tag{5}$$

$$Q_{evap-loss} = L_v h_c (P_1 - P_a)/(1.6 C_s P_{atm}) \tag{6}$$

$$Q_{HR} = \dot{m} \times c_p \times \left(T_{LCZ} - T_{inwf}\right) \times t \tag{7}$$

Efficiency of Solar Pond:

$$\eta_{SP} = \frac{Solar\ Radiation\ input\ in\ LCZ - Total\ Losses - Heat\ Removal\ from\ LCZ}{Solar\ Radiation\ input\ at\ surface\ of\ the\ solar\ pond} = Q_I - \left(Q_{hlg} + Q_{conv} + Q_{rad} + Q_{evap}\right) - Q_{HR})/Q_{I=0} \tag{8}$$

6. Experimental techniques for heat transfer analysis

When it comes to thermal energy storage, solar ponds are an exciting new development because of their low cost, long lifespan, and high efficiency. Accurate measurements of heat transmission within a solar pond could improve its design and operation. There are now experimental methods for measuring multiple properties associated with heat transport in solar ponds. This chapter provides a summary of the experimental methodologies for heat transfer analysis in solar ponds, including the measurement of temperature, salinity, velocity, and solar radiation.

6.1 Thermometer readings

Temperature readings are essential for characterizing the thermal behavior of a solar pond. Several techniques have been developed for gauging the temperature of a sun pond, including the use of thermocouples, resistance temperature detectors (RTDs), and infrared thermography. Most of the solar pond research utilizes thermocouples or resistance temperature detectors (RTDs) to track water temperatures. Thermocouples can swiftly and precisely record temperatures that are exceedingly high. RTDs can be used to keep tabs on temperatures for far longer than thermocouples can. Infrared thermography, a noncontact technique, can be used to create temperature maps of the pond's surface [21].

6.2 Quantifying salt content

Salinity measurements are one way to get a sense of the density gradient in a solar pond. The density gradient impacts both the storage capacity of the pond and the rate of convective heat transfer. The salinity of water in a solar pond can be measured in several ways, including by electrical conductivity testing and refractometry. Electrical conductivity is commonly used to determine the salinity of solar ponds. A simple and trustworthy method for determining the salinity of water is to measure its electrical conductivity.

6.3 Speed calculation

In a solar pond, the rate of convective heat transfer can be determined by monitoring the water's velocity. Several techniques have been developed for measuring velocities in solar ponds, including laser Doppler velocimetry (LDV), particle image velocimetry (PIV), and ultrasonic Doppler velocimetry (UDV). Laser Doppler velocimetry (LDV) and particle image velocimetry (PIV) are two optical methods for measuring water particle speeds. One noninvasive technique for measuring water particle velocity is ultrasonic Doppler velocimetry (UDV) [22].

6.4 Measuring solar radiation

The amount of solar radiation entering a solar pond can be used to characterize its energy input. Spectroradiometers, pyranometers, and other similar instruments can all be used to measure the sun's rays. The amount of solar radiation that strikes a surface can be calculated with the help of a pyranometer. A pyrheliometer is a device used to measure the incidence of solar flux on a surface. Instruments called spectroradiometers are employed in solar spectral analysis [23].

There are now experimental methods for measuring multiple properties associated with heat transport in solar ponds. The thermal behavior of a solar pond can be described by measuring its temperature, salinity, velocity, and solar radiation. Accurate measurements of these elements can enhance the pond's design and maintenance. Future experimental study on heat transfer in solar ponds can build on the foundation provided in this chapter.

7. Data analysis techniques that are commonly used in heat transfer analysis include

Data analysis is essential for heat transfer studies in solar ponds. Various statistical methods can be used to infer the thermal behavior of the solar pond after scientists have collected data through tests. Commonly used techniques will be discussed, such as regression analysis, principal component analysis, and neural networks.

Regression analysis is a common tool used by statisticians to explore the relationships between sets of data points. Heat transfer analysis for solar ponds can make use of several input characteristics, such as solar radiation intensity, air temperature, water depth, and heat transfer coefficient. To analyze the relationship between these inputs, a regression analysis can be carried out. A model of the solar pond's thermal behavior will be informed by these measurements. Several solar pond studies have employed regression analysis to better understand the phenomenon. A model for forecasting the hourly water temperature of a solar pond was developed using regression analysis [24]. They used the levels of solar radiation, air temperature, wind speed, and relative humidity to build their model.

Principal component analysis (PCA) is a type of multivariate data analysis that is used for this purpose. Principal component analysis allows us to determine which variables have the most effect on the temperature distribution in the solar pond. By using principal component analysis (PCA), the original variables are "transformed" into a new set of variables. These independent primary components explain a considerable deal of the natural variation in the original data. PCA has been employed in several solar pond studies. Using PCA [25], identified the most important contributors to a solar pond's thermal efficiency. They determined that the most crucial elements were pond depth, water salinity, and available sunlight.

Predictions of the thermal behavior of a solar pond can be made using machine learning methods like neural networks. Like the human brain, neural networks can make predictions based on past data. Heat transfer research on solar ponds can benefit from the use of neural networks to develop predictive models of water temperature. In their research on solar ponds, several experts have turned to neural networks. Using a neural network, for instance, researchers were able to predict [26] the temperature distribution in a solar pond. They used the amount of solar radiation, the air temperature, the wind speed, and the relative humidity as inputs to train their neural network.

In conclusion, data analysis methodologies are needed for heat transfer study of solar ponds. Regression analysis, principal component analysis, and neural networks are all prominent methods for dissecting data. Using these techniques, one may identify the most influential variables in the solar pond's thermal regime and develop predictive models for the pond's water temperature.

8. Validation of mathematical models for heat transfer analysis

It is common practice to use mathematical models to investigate the mechanisms of heat transport in solar ponds. Predicting the behavior of heat transfer effectively, however, requires first validating these models. This chapter explores the many methods for testing the accuracy of mathematical models of heat transfer in solar ponds. The first part of this chapter provides an overview of the heat-transfer governing equations that are relevant to solar ponds. Following that, a wide variety of validation techniques, from analytical validation to numerical validation to experimental validation, are covered. This chapter also highlights the need of sensitivity analysis in ensuring the accuracy of mathematical models. Finally, examples of possible applications of various validation strategies are provided in the form of case studies.

Mathematical models can be used to examine the dynamics of heat transfer in solar ponds. These models can be used to construct and optimize solar ponds for maximum efficiency and effectiveness. These models, however, are only as accurate as their capacity to reproduce the physical characteristics of a solar pond. This means that checking the models' predictions against experimental data is essential for ensuring reliability. In this chapter, we will go over a variety of techniques for checking the validity of mathematical models used to predict heat transfer in solar ponds.

The continuity equation, the momentum equation, the energy equation, and the salt concentration equation are the governing equations for heat transmission in a solar pond [27]. Mass conservation is expressed by the continuity equation, which is:

$$\nabla.(\rho v) = 0 \tag{9}$$

where the fluid density, the velocity vector v, and the divergence operator all play important roles. Momentum conservation is represented by the following equation:

$$\rho(v.\nabla)v = -\nabla P + \rho g + \mu\nabla^2 v \tag{10}$$

where P represents pressure, g represents gravity, u represents fluid viscosity, and 2 represents the Laplacian operator. Conservation of energy is expressed by the equation:

$$k^2 T + Q = c_p\left(\frac{T}{t} + vT\right) \tag{11}$$

The temperature T, the thermal conductivity k, and the quantity Q represent the heat source; where Cp is the specific heat capacity. The equation for the concentration of salt, which stands for the principle of salt conservation, is as follows:

$$\left(\frac{c_2}{t_2} + \frac{v_2}{c}\right) = D^2 c \tag{12}$$

D = Diffusion Coefficient, where c = Salt Concentration.

By comparing the analytical answers obtained by solving the mathematical model with the analytical solutions obtained by solving simplified models, we may do analytical validation, the simplest validation technique. Analytical solutions, which are precise solutions obtained for boundary conditions, can be used to check whether the model is true [28].

Numerical validation involves comparing the predicted results of a mathematical model to observed data. Numerical methods such as finite element analysis, finite difference analysis, and boundary element analysis are used to solve the governing equations. Numerical solutions are compared to experimental data to see if the model is accurate.

Experimental validation involves checking the model's predictions against data obtained from a functioning solar pond. The temperature, salinity, and velocity profiles of the water in the solar pond can be mapped using sensors like thermocouples, conductivity meters, and flow meters. The model is validated by comparing its predictions to data collected in experiments.

Sensitivity analysis, which is an essential aspect of validation, examines how the model's predictions change when the input parameters are varied. By identifying the characteristics that have a significant effect on the model's predictions, sensitivity analysis helps guide experimental design and data collecting [29].

Examples of Use: We illustrate various validation techniques with real-world situations. These experiments are used to check the accuracy of mathematical models of solar ponds with different shapes, salinity gradients, and heat exchangers. Validating mathematical models for heat transfer analysis in solar ponds relies heavily on boundary conditions. Boundary conditions are crucial to the accuracy of the mathematical model. It is essential that the boundary conditions properly reflect the actual setting of the solar pond. The surface temperature of the pond, for instance, will be influenced by factors such as the surrounding air temperature and solar radiation. Heat transfer at the pond's base is influenced by the temperature of the ground below.

Experimental data can be used to help validate mathematical models used to analyze heat transmission in solar ponds. Several experimental methods, such as thermal imaging, temperature probes, and heat flow sensors, can collect data for validation. Validity of the model can be determined by checking its predictions against actual data [30].

Mathematical models employed in heat transfer research on solar ponds can also be evaluated by comparing their predicted findings to those of other models. This technique is known as model benchmarking. Benchmarking models can be useful when there is either a dearth of experimental data or insufficient data to do a full validation. This technique compares the model's predictions to those of other, credible models to determine how well it performs.

It has been determined that mathematical models are a useful tool for analyzing heat transmission in solar ponds. However, validation of the models is required to guarantee that they accurately depict the physical reality of the solar pond. Validation can be achieved by comparing the model's predictions with either experimental data or the predictions of other validated models. Better solar ponds can be designed and run if the models used for heat transfer analyses are validated [31].

9. Optimization of solar pond performance for heat transfer analysis

Low-temperature heat can be stored in solar ponds, which are simple and affordable to build. However, the efficiency of a solar pond can be improved by optimizing its design and operations. This chapter will examine heat transport analyses relevant to the design and operation of solar ponds.

9.1 Construction optimization

During the planning phase of solar ponds, the optimal pond geometry and material characteristics should be selected. In order to enhance solar radiation absorption and reduce heat loss by convection, radiation, and conduction, the design of the pond should be selected. The shape of a solar pond is often a circle or a rectangle. Features of the liner and insulation layer can also affect how well the pond functions. In addition to being stable in the presence of high salt concentrations and direct sunshine, the ideal liner material would also have a low thermal conductivity. The thermal resistance of the insulation layer should be high to minimize heat loss to the ground [32].

9.2 Optimization of procedures

Several operational parameters, including salinity gradient, water depth, and water temperature, can be modified to increase solar pond productivity. The salinity gradient should be chosen to produce a substantial density differential between the layers to optimize convective heat transmission. The water level in the pond needs to be regulated so that it can store as much heat as possible while allowing the least amount of heat to escape through convection and radiation. Liner material degradation and heat loss by radiation can be prevented by maintaining water temperatures within the working range [33].

9.3 Optimization theory

More effective solar pond designs and operations are possible with the use of mathematical optimization techniques. These strategies employ mathematical models to anticipate the pond's behavior in certain situations. The models can be used to do a sensitivity analysis, revealing which factors have the greatest bearing on the pond's performance. Using the models, optimization studies can be conducted to find the optimal pond configuration and operating parameters [34].

Optimizing heat transfer analysis is a crucial aspect of designing and operating a solar pond. Design optimization involves picking the optimal pond shape and materials, whereas operation optimization involves adjusting operating settings. Mathematical optimization methods can be used to enhance solar pond design and operation by calculating pond performance under different design and operating scenarios. By optimizing performance, solar ponds can become more efficient and cost-effective.

10. Applications of solar ponds relevant to heat transfer analysis

Solar ponds are a one-of-a-kind thermal energy storage technique that can collect the sun's heat and keep it in reserve for later use. Due to the many layers of water with varying salt concentrations, natural convection currents are used to transport heat from the surface to the depths in these systems. As the world searches for more sustainable energy sources, solar ponds have become increasingly popular for storing heat. In this section, we will examine the usage of solar ponds in heat-transfer studies.

Solar ponds have several applications, but one of the most prevalent is in agricultural settings. Solar ponds can offer the constant heat needed for greenhouse horticulture to maintain optimal growing conditions. A solar pond can be used to heat the greenhouse at night or on overcast days, when sunshine is scarce. Many countries use crop drying as a routine agricultural activity, and solar ponds can be used for this. The crop drying process can be sped up and energy costs reduced by using the pond's thermal energy to power a crop dryer [35].

Many manufacturing procedures can benefit from the consistent heat provided by solar ponds. For instance, solar ponds can be used for desalination in places where freshwater is in short supply. Utilizing the pond's thermal energy to power a desalination plant is a great way to cut down on the price of desalination. Using solar ponds for heating and cooling can reduce the amount of energy used by traditional HVAC systems in industrial buildings [36].

One of the many residential applications for solar ponds is the production of hot water for household consumption. The thermal energy from the pond can be used to heat the water in the tank, which can then be used for bathing, cooking, and other household tasks requiring hot water. Solar ponds are an eco-friendly alternative to traditional heating systems for swimming pools [37].

Solar ponds, in conclusion, are a novel and adaptable thermal energy storage system with several applications in the study of heat transfer. Some examples of greenhouse, industrial, and domestic applications are desalination, space heating and cooling, hot water generating, and swimming pool heating. As the world searches for more sustainable energy sources, solar ponds have become increasingly popular for storing heat. As technology improves, solar ponds will become an even more attractive option for sustainable energy generation [38].

11. Case studies relevant to heat transfer analysis

Solar ponds have been the subject of extensive research due to their promising potential as a means of storing and harnessing solar energy for a broad variety of applications. Heat transport analysis of solar ponds has been the subject of extensive experimental and numerical study. The practical application of heat transfer analysis in the design and optimization of solar ponds is explored in this chapter.

Researchers in Iran investigated the feasibility of using a solar pond to provide continuous greenhouse heating. The solar pond was to be 1.5 meters deep, with a salinity gradient of 80 g/L, and cover 100 square meters of ground. Finite element analysis was used to determine the amount of heat lost by the pond. The results showed that the greenhouse could be kept at a comfortable 20–25 degrees Celsius with heat from the pond for 12 hours every day. Greenhouses may be heated effectively and sustainably with a solar pond, it was found [39].

A solar pond desalination system was the focus of a study in Saudi Arabia. The solar pond had an area of 1400 m^2, was 4 meters deep, and had a salinity gradient of 110 g/L. Finite element analysis was used to determine the amount of heat lost by the pond. The results showed that a temperature difference of 30 degrees Celsius could be maintained between the pond's upper and lower levels. A multi-effect distillation device powered by the temperature difference produced 300 cubic meters of clean water every day. It has been found that desalinating saltwater in a solar pond is a viable and eco-friendly solution [40].

In an Egyptian study, researchers looked at the feasibility of using a solar pond for fish farming. A 500-square-meter solar pond with a 3-meter depth and an 80-grams-per-liter (g/L) salinity gradient was designed. Finite element analysis was used to determine the amount of heat lost by the pond. The results showed that the pond maintained a constant temperature of 30 degrees Celsius, which is perfect for the growth of tilapia. The solar pond was found to be a practical and low-cost method of establishing an aquaculture-friendly ecosystem [41].

The case studies presented in this chapter show how heat transfer analysis can be used to optimize the design of solar ponds. Solar ponds can be used to desalinate water, produce fish for human consumption, and heat water. Numerical tools, such as finite element methods, have been utilized for analysis and enhancement of solar ponds. This research shows that solar ponds can be used to solve the problem of energy and water scarcity in a variety of settings.

12. Conclusion and future directions

In conclusion, solar ponds are an exciting new method of collecting solar energy for use in a wide range of industries, from heating and cooling to desalinization and power generation. This chapter has emphasized the significance of heat transfer analysis in solar pond design and operation. Solar ponds store solar energy by thermal stratification, and the resulting hot water or steam can be used for a wide range of purposes. Predicting solar pond performance and making design and operating adjustments based on heat transfer studies are crucial. The rate at which heat is transferred from a solar pond to the surrounding environment is dependent on several variables. Insulation, redesigning the solar collector, and installing heat exchangers are only some of the methods that can be utilized to boost the efficiency of solar ponds.

There are several restrictions and questions that need to be answered about solar ponds, despite their potential advantages. One disadvantage of solar ponds is the expensive initial investment and ongoing maintenance costs associated with them. Additionally, more study is required to examine the environmental impact of employing large-scale solar ponds, as well as the long-term performance of solar ponds under various operating situations.

Improving the efficiency and lowering the cost of solar ponds, as well as discovering new uses for this technology, should be the primary goals of future research in this subject. Furthermore, new and improved solar pond designs may be possible in the future thanks to developments in materials science and technology.

In conclusion, solar ponds present an attractive strategy for making use of renewable energy sources and satisfying the rising worldwide need for energy. To overcome the technological and economic hurdles of this technology, however, more study is required. Solar ponds have the potential to reduce energy costs and carbon emissions, but more study is needed to determine how best to construct and operate them for different uses.

Author details

Baljit Singh Bhathal Singh
School of Mechanical Engineering, College of Engineering, Universiti Teknologi MARA, Selangor, Malaysia

*Address all correspondence to: baljit@uitm.edu.my

References

[1] Bisquert J, Pons S, Guilleminot J. Modeling and Simulation of Solar Ponds. Springer Science & Business Media; 2010

[2] Reynolds JH. The solar pond: A simple, efficient solar energy collector with widespread applications. Applied Energy. 1981;**8**(3):165-190

[3] Tiwari GN, Rahim NA, Dubey S. Solar Ponds for Renewable Energy Generation: Efficiency and Applications. CRC Press; 2018

[4] Garg HP. Solar Energy: Fundamentals, Design, Modelling and Applications. Elsevier; 2016

[5] Madhlopa A. Experimental investigation of heat transfer in a solar pond. Journal of Solar Energy Engineering. 2015;**137**(1):011008. DOI: 10.1115/1.4028491

[6] Hepbasli A. A key review on exergetic analysis and assessment of renewable energy resources for a sustainable future. Renewable and Sustainable Energy Reviews. 2006;**10**(6):528-555. DOI: 10.1016/j.rser.2004.09.003

[7] Kalogirou S. Solar Energy Engineering: Processes and Systems. Elsevier; 2009

[8] Sharma S, Sharma S. Modeling and optimization of solar pond: A comprehensive review. Solar Energy. 2019;**191**:196-218. DOI: 10.1016/j.solener.2019.09.014

[9] Hu J et al. Experimental investigation of a salt-gradient solar pond for low-grade heat collection and storage. Applied Thermal Engineering. 2018;**128**: 1005-1016. DOI: 10.1016/j.applthermaleng.2017.09.101

[10] Hossain S et al. Experimental investigation of solar pond performance for thermal energy storage. Energy Conversion and Management. 2021;**243**: 114401. DOI: 10.1016/j.enconman.2021.114401

[11] Hu J et al. A comprehensive review on solar pond technologies: Design, performance, and applications. Renewable and Sustainable Energy Reviews. 2018;**82**(Part 3):2973-2993. DOI: 10.1016/j.rser.2017.11.064

[12] Chandrasekhar B et al. Performance evaluation of a hybrid solar pond integrated with heat pipe based low temperature desalination system. Desalination. 2021;**505**:114916. DOI: 10.1016/j.desal.2021.114916

[13] Al-Karaghouli A, Kazmerski LL. Energy and exergy analyses of a salt-gradient solar pond. Energy Conversion and Management. 2012;**58**:10-20. DOI: 10.1016/j.enconman.2011.12.023

[14] Chandra R, Tiwari GN. Performance evaluation of a salt gradient solar pond integrated with solar still. Desalination. 2005;**175**(1):11-21. DOI: 10.1016/j.desal.2004.08.013

[15] Tiwari GN, Chandra R, Ghosal MK. Design, performance and applications of solar ponds: A review. Renewable and Sustainable Energy Reviews. 2007;**11**(5): 713-728. DOI: 10.1016/j.rser.2005.07.005

[16] Jafarkazemi F. A review on solar pond applications: Energy and water desalination. International Journal of Environmental Science and Technology. 2019;**16**(7):3547-3562. DOI: 10.1007/s13762-018-2134-1

[17] Dehghan Manshadi H. Performance evaluation of solar pond as a heat source

in a solar-driven multigeneration system. Renewable Energy. 2019;**138**:284-296. DOI: 10.1016/j.renene.2019.02.047

[18] Kumar A, Rosen MA. Numerical simulation of solar pond performance for thermal energy storage. Energy Conversion and Management. 2017;**141**: 43-52

[19] Al-Sulaiman FA, Zubair SM. Numerical study of heat transfer in solar pond with different configurations. Journal of Thermal Analysis and Calorimetry. 2017;**129**(2):1099-1109

[20] Sharma RK, Tyagi VV, Chen CR, Buddhi D. Review on thermal energy storage with phase change materials and applications. Renewable and Sustainable Energy Reviews. 2009;**13**(2):318-345

[21] Chandrasekar S, Edwin Geo V. A review on experimental investigations of solar ponds. Renewable and Sustainable Energy Reviews. 2016;**62**:740-758

[22] Hachicha AA, Hamzaoui AH, Nasrallah SB. Experimental study of the convective and radiative heat transfer in a solar pond. Energy Procedia. 2016;**74**: 784-793

[23] Yan Z, Yang H, Chen Z. Recent advances in measurement techniques for solar radiation. Renewable and Sustainable Energy Reviews. 2013;**27**:573-586

[24] El-Sebaii AA, Al-Ghamdi AA, Al-Hazmi FS. Modeling of hourly water temperatures of a solar pond using artificial neural networks and regression analysis. Renewable Energy. 2008;**33**(3): 491-500

[25] Bilgen E, Keles S, Cuce E. Analysis of thermal performance parameters of a solar pond using principal component analysis. Energy Conversion and Management. 2003;**44**(20):3295-3311

[26] Boukhris M, Bouaichaoui S, Nasri M. Predicting the temperature distribution in a solar pond using artificial neural networks. Energy Conversion and Management. 2013;**75**:299-306

[27] Akbarzadeh A et al. A review on mathematical models of solar ponds. Renewable and Sustainable Energy Reviews. 2016;**57**:850-866

[28] Bhattacharya SK et al. Mathematical modeling of solar ponds: A review. Renewable and Sustainable Energy Reviews. 2014;**30**:550-561

[29] Duffie JA, Beckman WA. Solar Engineering of Thermal Processes. John Wiley & Sons; 2013

[30] Fudholi A et al. A review of solar ponds for sustainable energy development. Renewable and Sustainable Energy Reviews. 2017;**70**:1147-1160

[31] Kumar S et al. Mathematical modeling of heat transfer in solar ponds: A review. Renewable and Sustainable Energy Reviews. 2016;**55**:907-922

[32] Kalogirou SA. Solar Energy Engineering: Processes and Systems. Academic Press; 2009

[33] Tiwari GN, Ghosal MK. Advances in Solar Energy Technology. Alpha Science International Ltd.; 2005

[34] Reddy TA, Kaushika ND. Solar Ponds: Fundamentals, Technology and Applications. TERI Press; 2011

[35] Krüger E, Gabbasa M, Ochieng RM. Solar ponds: A review. Renewable and Sustainable Energy Reviews. 2019;**101**: 631-648

[36] Solanki SC. Solar Energy: Fundamentals, Design, Modeling and

Applications. PHI Learning Pvt. Ltd; 2010

[37] Akbarzadeh A, Sheikhzadeh GA, Kianifar A. Solar ponds: Principles, design, and applications. Renewable and Sustainable Energy Reviews. 2013;**28**: 494-506

[38] Sarbu I, Sebarchievici C. Solar energy storage using phase change materials and solar ponds: A review. Renewable and Sustainable Energy Reviews. 2017;**79**:1211-1227

[39] Esfahani JA et al. Design, construction and evaluation of a solar pond for heating greenhouses in Iran. Energy Conversion and Management. 2017;**142**:334-341

[40] Al-Karaghouli A et al. Performance of a solar pond for seawater desalination under Saudi Arabian conditions. Desalination. 2010;**252**:70-76

[41] El-Naggar ME et al. Performance of a solar pond for aquaculture purposes in Egypt. Desalination and Water Treatment. 2016;**57**:17767-17773

Chapter 8

Perspective Chapter: Smart Liquid Cooling Solutions for Advanced Microelectronic Systems

Montse Vilarrubí

Abstract

Thermal management is today a primary focus in the electronics industry due to the continuous increase of power density in chips increasingly smaller in size, which has become a critical issue in fast-growing industries such as data centers. As air-cooling fails to meet the high heat extraction demands of this sector, liquid cooling emerges as a promising alternative. Nevertheless, advanced microelectronic components require a cooling system that not only reduces the energetic consumption but also enhances the thermal performance by minimizing the thermal resistance and ensuring high-temperature uniformities, especially under variable heat load scenarios with high heat dissipating hotspot regions, where conventional liquid cooling solutions prove inefficient. This chapter provides an overview of different passive heat transfer enhancement techniques of micro heat sinks from the literature, focusing on intelligent and adaptive solutions designed to optimize the cooling performance based on local and instantaneous cooling requirements for non-uniform and time-dependent power distribution maps.

Keywords: liquid cooling, electronics thermal management, adaptive cooling, heat transfer enhancement, micro heat sinks, variable hotspots, energy efficiency

1. Introduction

Moore's law, formulated in 1975, projected that the number of transistors in a compact integrated circuit (IC) would double every 24 months. Over the past few decades, the progress in circuit density has closely matched this prediction [1]. At the same time, Dennard scaling law stated that as transistors shrink in size, their power density remains constant. However, in recent years, this law seems broken as the performance enhancement slowed down while the number of transistors per circuit is still increasing [2]. As a result, the continuous increment in power density of ever smaller ICs has led to an exponential rise in heat dissipation, and thermal management emerged as a primary concern for the industry [3].

Additionally, the escalating demand for data processing, networking, or data storage systems has led to a big expansion of the data center industry, which presently contributes to 1–1.5% of global electricity consumption, with CO_2 emissions equivalent to those generated by the airline industry and has an estimated growth of 500% by 2030 [4–6].

IntechOpen

To ensure the temperature of the IT equipment remains within its safe operational range, data centers need effective cooling mechanisms in place, which traditionally involved the use of large cooling Heating, Ventilation, and Air Conditioning (HVAC) systems. Then, within a data center facility, the cooling system stands out as one of the most energy-intensive components, accounting for 30–40% of the total energy consumption [7] (**Figure 1**).

Until recently, air-based solutions, including passive methods relying on natural convection or active cooling involving fans or heat sinks, were extensively employed for electronic component cooling. However, with the increasing heat dissipation demands of advanced microelectronic systems, conventional air cooling mechanisms are insufficient to provide the needed heat removal capacity, and water-based systems become preferred [9, 10].

Since Tuckerman and Pease [11] demonstrated the viability of microchannels liquid cooling for electronic chips back in 1981, this technology has been subject to extensive research to improve its performance, including geometric modifications [12, 13], the use of nanofluids or the study of alternative technologies such as heat pipes, jet impingement, or spray cooling [14, 15]. Although this technology achieved low thermal resistances, it still has the challenges of poor temperature uniformities, which can result in serious damage to the performance and reliability of the chips [16], and large pressure drops, which result in more energy-demanding cooling devices due to higher pumping power.

Advanced microelectronic components, such as multicore processors or three-dimensional integrated circuits (3D-ICs) are currently being persuaded by the IC industry due to their high performance and low power consumption [17]. Multicore processors are based on multiple independent execution cores on a single CPU that can work in parallel tasks, where each core dissipates multiple times the heat flux dissipated at the rest of the chip, which leads to the appearance of hotspot regions and non-uniform power map distributions [18–20]. Similarly, 3D-ICs architectures stack multiple dies, interconnected through vertical connectors to extend the performance

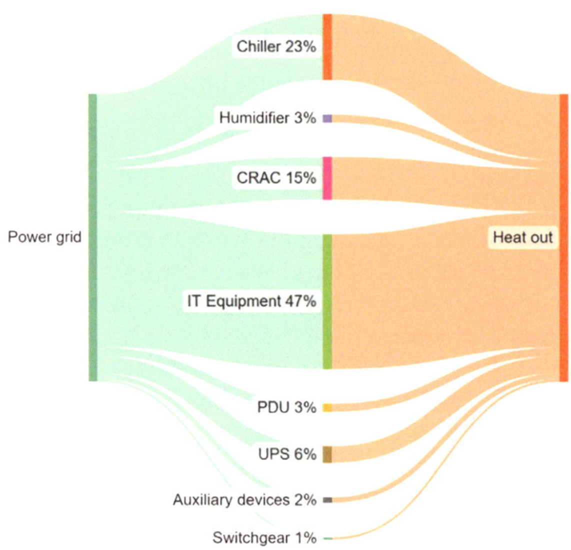

Figure 1.
Sankey diagram showing the average power consumption of the different components of a data center [8].

of 2D chips [9]. Here, thermal management becomes critical as the heat-generating layers are aligned in adjacent films creating localized hotspots, and any dissimilarity between power maps of individual chips constituting the stack would add another geometrical dimension to the thermal non-uniformity [21, 22].

Hotspot regions can dissipate multiple times the heat flux dissipated at the rest of the chip, requiring high flow rates for effective cooling. Nevertheless, as these regions are typically localized, maintaining constant large flow rates across all cooled devices can result in overcooled systems. Various attempts implementing single-phase liquid cooling have been made to redirect the coolant to the most demanding zones while reducing the required pumping power [23–25]. For example, some authors proposed the use of a cold plate with variable microchannel width or pin-fin density to effectively dissipate high heat fluxes from small areas and achieved reduced thermal gradients at the package level with low-pressure drop, especially when combining this system with a central inlet and two side outlets [26–30]. In a different line, Lu et al. [31] proposed the use of fixed vortex generators along the height of a rectangular microchannel, placed upstream of the hotspot region, and their numerical results indicated that the adoption of these flow disruptive elements can significantly improve the cooling effect over the hotspot with a lower pressure loss penalty compared with plain channels. Other authors followed the same research line with different degrees of success [32, 33]. However, all previous detailed systems rely on fixed geometries and lack the ability to adapt their behavior to changing heat load scenarios across time and space, so these systems are only optimized for specific conditions, leading to oversized pumping powers and poor temperature uniformities for varying operational parameters.

Consequently, advanced cooling solutions that aim to improve the performance of current microelectronics systems should focus on devices capable of dynamically adapting to changing boundary conditions across time and space. These solutions should simultaneously strive for reduced energy consumption and improved thermal performance, with reduced thermal resistances and high-temperature uniformity.

2. Overview of passive heat transfer enhancement techniques in micro heat sinks (MHS)

Different heat transfer enhancement techniques have been explored in the literature to improve the cooling capabilities of microscale heat exchangers and overcome their limitations. Tao et al. [34] identified three different mechanisms to improve the thermal transfer: reducing the thermal boundary layer, introducing flow interruptions along the flow path, and enhancing the velocity gradient near the heated surface. Among the common passive heat transfer enhancement methods, which do not require external energy use, are channel shape modifications, surface roughness adjustments, incorporation of fluid additives, and the addition of flow disruption elements inside the channel [35, 36].

2.1 Modification of channel shape

Several researchers have demonstrated that curved flow paths can enhance heat transfer by generating secondary flows and Dean vortices [37]. Accordingly, various studies have evaluated the thermal performance improvements of the microchannel heat sink (MCHS) with curved walls, concluding that the generation of vortices in the channel cross-sections promotes convective heat transfer [38–40] (**Figure 2**).

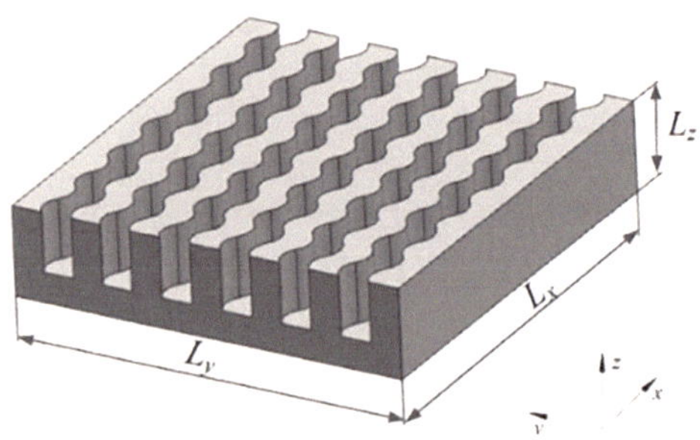

Figure 2.
Schematic of a wavy microchannel heat sink [39].

Additionally, the incorporation of superhydrophobic walls has shown significant potential to reduce the pressure drop of the cooling devices [41].

The effect of cross-sectional channel aspect ratio and shape have been found to have a significant influence on the heat transfer and fluid flow characteristics of the MCHS [42, 43]. For example, Wang et al. [44] investigated the effects of different channel shapes on the heat transfer and fluid flow characteristics of a microchannel and observed that rectangle channels exhibited better overall performance than other shapes but also had lower thermal resistances. Other researchers have focused on the generation of secondary flows, which can lead to significant thermal enhancement but with an added pressure drop penalty [45, 46]. However, the combination of ribs and secondary channels has been shown to further enhance mixing while reducing pressure drop [47]. In a different line, various researchers have evaluated the heat transfer enhancement in double-layered microchannel heat sinks (DL-MCHS), which exhibited lower temperature rise and pressure drop compared to conventional MCHS [48–50] (**Figure 3**). Also, the use of porous surfaces attracted the attention of researchers to improve thermal performance with reduced pressure drops, both in single-layer and double-layer MCHS [52, 53].

2.2 Surface roughness

Another passive thermal enhancement technique discussed in the literature involves modifying the characteristics of the heated surface through variations in surface roughness. This approach has been found to increase the flow resistance and improve heat transfer [54–57]. For instance, Gamrat et al. [54] presented a model

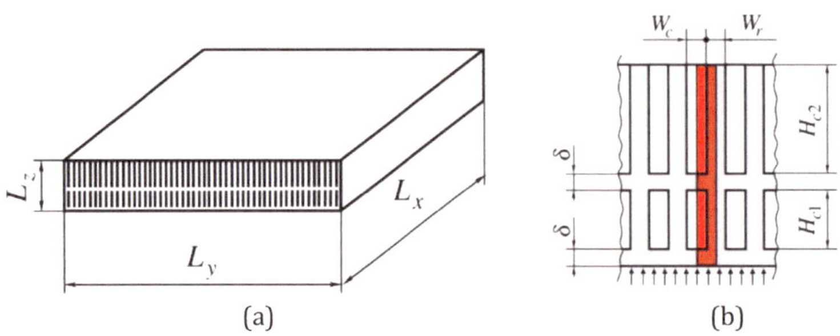

Figure 3.
Parallel double-layered microchannel heat sink: 3D-view (a) and cross-section view (b) [51].

in which roughness was modeled as a pattern of parallelepiped elements distributed periodically within microchannels. They studied the influence of roughness on heat transfer in laminar flows and introduces a relative roughness value (k*) able to provide thermal performance values higher than one. More recently, Madev and Manay [57] conducted experimental evaluations to study the effect of induced surface roughness on multiple stainless-steel microchannels and revealed a notable impact on mixed convective heat transfer, which decreased with the augmentation of the hydraulic diameter of the channel.

2.3 Fluid additives

The addition of solid nanoparticles into the working fluid has been investigated by different researchers as a promising method to enhance heat transfer [58, 59]. Also, various studies explored the combination of nanofluids with improvements in geometrical parameters of the heat sink, such as the use of vortex generators [60], ribs and grooves [61, 62], pin fins [63], or jet impingement [64]. For example, Heydari et al. [61] assessed the effect of combining triangular ribs inside an MCHS with different nanofluids and concluded that an increase in the volume fraction of nanoparticles, as well as the use of nanoparticles with smaller diameters, lead to greater heat transfer. The authors also observed that the friction coefficient and pumping power remained nearly independent of nanoparticle diameter. On the other hand, Alkasmoul et al. [65] evaluated the thermohydraulic performance and feasibility of various nanofluid types and concentrations in an MCHS. Their results stated that the effect of increasing flow rate had a more dominant effect than increasing nanoparticle concentrations, while at the same time, all nanofluids required higher pumping power than pure water at all concentrations (**Figure 4**).

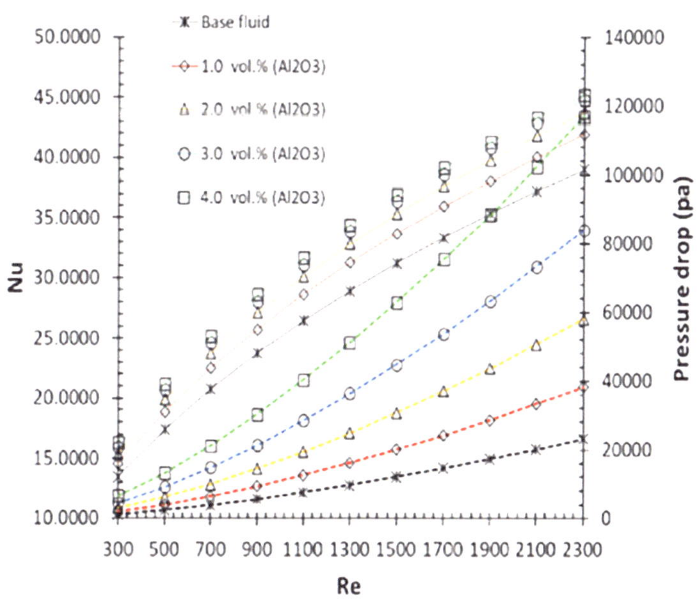

Figure 4.
Nusselt number (solid line) and pressure drop (dashed line) in an MCHS at different Reynolds numbers for water and various concentrations of nanofluid [65].

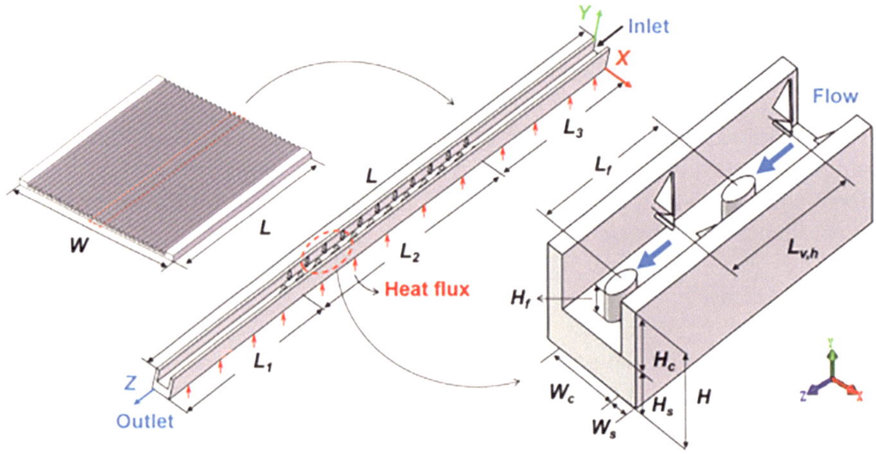

Figure 5.
Schematic diagram of an microchannel heat sink with pin fins and vortex generators acting as flow-disturbing elements.

2.4 Flow disruption

A widely employed technique for thermal enhancement involves the incorporation of flow interruptions within the channel flow. These interruptions disrupt the flow field and the growth of the boundary layer and induce fluid swirling along the cooling channel, leading to better mixing and redevelopment of this layer, which results in lower thermal resistance [66]. The different disturbing elements used for heat transfer enhancement are based on sidewall obstructions, such as ribs and cavities [67], or flow obstructions like wings, winglets [68], ribs [69], pin fins [70], or surface protrusions (**Figure** 5) [71]. However, the overall performance enhancement of an MHS when using VGs depends on many geometric parameters such as their shape, dimensions, position, or angle of attack [60].

2.5 Current limitations

Although passive heat transfer techniques have been demonstrated to provide extended cooling capacity to microscale heat exchangers, they also induce additional hydraulic resistance at the flow pass, which leads to higher pressure drops and pumping power on a continuous basis. Furthermore, these systems are typically designed to handle the worst-case situation and lack adaptability to varying time-dependent and non-uniform power distribution scenarios, which leads to overcooling when the heat extraction demands are not maximized. To address these challenges, there is a need for more intelligent and adaptive cooling solutions that can dynamically respond to changing heat load conditions and optimize cooling performance while minimizing unnecessary energy consumption.

3. Smart heat transfer enhancement techniques in MHS

In last years, smart cooling solutions started to focus on the performance optimization of MHS when submitted to changing boundary conditions in time and space, as the ones induced by advanced electronic systems such as multicore processors, SoC, or 3D-ICs. In these cases, the cooling systems should be able to maximize their thermal performance to high heat extraction demands at peak points while minimizing the energetic consumption when the cooling demands decrease. Therefore, low thermal resistances, high-temperature uniformities, and reduced pumping power are the main goals to achieve under variable power scenarios both in time and space.

3.1 Flow rate regulation

Different studies suggested employing different actuators as valves to enable local flow rate regulation at the different regions of the cold plate based on the instantaneous cooling requirements. This intelligent regulation allows the redirection of more flow to the hotspot regions, achieving a more uniform temperature profile at the cooled device, effectively optimized for any variable thermal load scenario. However, the addition of valves inside the cold plate results in an increment of the pressure drop and so, higher energetic requirements when the valves are in the closed position.

For example, Azarkish et al. [72] analytically investigated the thermal response of temperature-regulated microvalves able to adapt the coolant mass flow rate distribution based on the local chip temperature. Linear and exponential responses of the microvalve were evaluated, obtaining in both cases a significant reduction in mass flow rate, while temperature uniformity across the chip was significantly improved with exponential microvalves. Similarly, Amnache et al. [73] detailed the fabrication process flow of Ag non-linear doubly clamped beams to implement an array of self-adaptive valves in a liquid-cooled heat sink. The fabrication was successfully done in a clean room through a lithography process. Following the same research line, Laguna et al. [74] experimentally evaluated the potential of a microfluidic cells cooling system, where each cell integrates a self-adaptive microvalve capable of tailoring the flow rate to the local and instantaneous cooling needs. The performance of this cooling system was assessed and compared with regular microchannels under variable heat loads. The results indicated a maximum temperature reduction of 15°C and a 74.7% decrease in pumping power when implementing the self-adaptive flow rate control. These findings highlight the potential of this cooling solution as a promising configuration for local and instantaneous thermal management control.

With a different working principle, other authors took advantage of the intrinsic shape memory effect of shape memory alloys (SMA) to develop thermally adaptive valves. Waddell et al. [75] employed SMAs as microfluidic valves inside a cooling channel to regulate the flow disturbance and enhance the local heat transfer performance and, although the combination of MCHS with SMA exhibits good performance in cooling hotspots, authors concluded that this solution could not be applied in practical large-scale industry due to its high cost. More recently, Vilarrubí et al. [76] studied the use of an array of bimorph SMA/metallic valves inside the cooling channels of a liquid-cooled cold plate designed for a rack server. The work demonstrated the capacity of these valves to double the flow rate at each channel for high cooling demands, resulting in energy savings of up to 30%. However, the authors also highlighted the challenges associated with this material due to its intrinsic thermal hysteresis cycle when operating outside its full phase transformation range, which limits the implementation of the solution (**Figure 6**).

Another approach of interest is the use of soft materials for a thermally responsive flow control, based on polymer gel-type materials. In this case, when the temperature of the hydrogel is lower than a determined temperature, the polymer chains inside

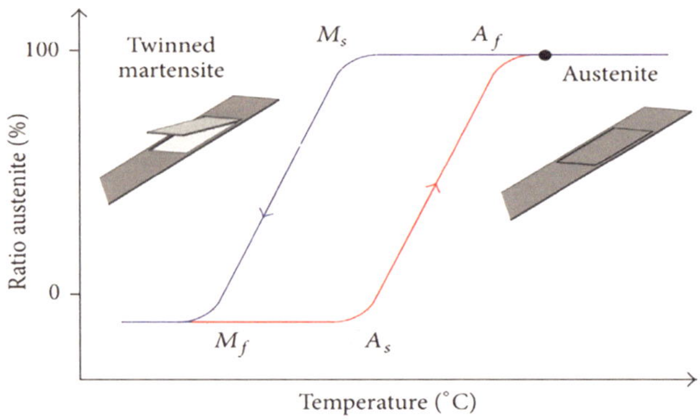

Figure 6.
Crystallographic behavior of SMA valves in function of its temperature. A_s refers to the starting temperature of the austenite phase transformation, A_f to the finishing temperature of the same phase, M_s is the starting temperature of the martensite phase and M_f the finishing temperature of the same phase transformation [76].

the hydrogel are hydrated and its volume increase dramatically. Contrarily, if the hydrogel temperature increases over a certain limit, the polymers begin to collapse and contract suddenly, so the volume decreases greatly in a short time. The main benefits of this technology include a quick thermal response with significant volume changes, easier fabrication with complex shapes, and low cost. Tudor et al. [77] reported the synthesis, characterization, and performance of thermo-responsive hydrogels as temperature-controlled actuators within microfluidic devices, where the hydrogel size was modulated by localized changes in its temperature due to the lower critical solution temperature behavior exhibited by the hydrogel. The same principle was studied by Yan et al. [78] to automatically regulate the mass flow rate distribution in a fractal microchannel heat sink when submitted to random hotspots. The authors stated that the swollen volume could reach 50–80 times the contracted volume and numerically demonstrated a reduction of 4.47°C in the maximum temperature rise of the heat sink when using hydrogels as valves. However, they observed that extreme hotspot conditions lead to worse uniform cooling as all hydrogels are in the same state. Also, Li and Xuan [79] proposed an embedded cooling system of a microchannel/pin fin heat sink with thermo-sensitive hydrogels located at the outlet channel. The obtained results demonstrated that heat flux of up to 500 W/cm^2 could be dissipated, with a pressure drop of 34.0 kPa and a reduction of peak temperature of 12.2°C, due to the studied self-adaptive cooling. The same authors also proposed the integration of hydrogel valves in silicon manifold channels and evaluated its impact on various hydrogel compositions. In this case, random hotspots of 460 W/cm^2 were effectively removed with a flow rate of 0.87 mL/s, and surface temperature uniformity was reduced to 27.2°C, achieving effective adaptive cooling of random hotspots [80].

Other works focused on developing external control mechanisms for flow rate adjustments, to match the instantaneous cooling requirements and achieve significant pumping power savings. For instance, Zhang et al. [81] developed an adaptive control law for the thermal-fluidic control of a microchannel evaporator, which estimated the heat transfer coefficient and adjusted the flow rate accordingly. Also, Da Luz et al. [82] proposed to adapt the pumping conditions to provide the minimum flow rate required to maintain the maximum chip temperature, through a variable flow miniaturized pump system with an associated electronics drive (**Figure 7**). Recently, Shahi et al. [83] explored the implementation of active flow control at the server level

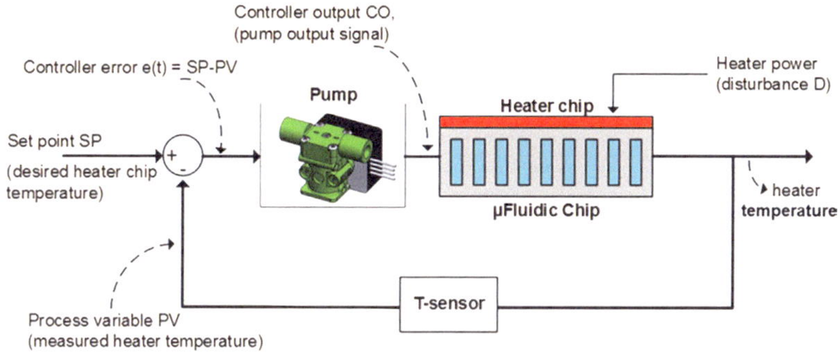

Figure 7.
Simplified schematic of the control loop block diagram for external control of the flow rate in function of the maximum cold plate temperature [82].

to regulate the coolant flow rates. They employed a ball valve connected to a micro engine to control the valve position via predefined rotational angles. While these strategies can adapt the overall flow rate of the cooled device, they may not efficiently regulate conditions for variable heat loads inside the same cold plate, which could turn in the overcooling of background areas and poor temperature uniformity. Thus, it could be a useful approach to combine with other adaptive solutions placed inside the cold plate.

3.2 Heat transfer regulation through flow disruption of fluid boundary layers

The effectiveness of different flow disruption techniques in enhancing the heat transfer within a MHS has been well established. However, their lack of adaptability to variable power scenarios results in oversized pumping powers and unnecessary energy consumption. To overcome this drawback, some researchers proposed to optimize the heat transfer capacity of the cooled device through the addition of smart mechanisms able to disrupt the fluid boundary layer and increase the mixing within the cooling channel when the applied heat loads vary in time and space. In these cases, the heat transfer enhancement capabilities of the flow-disturbing elements are only applied at high heat loads, remaining in a neutral position otherwise to minimize the pressure drop along the channels.

In 2001, Champagne and Bergles [84] introduced a novel concept involving a variable roughness heat exchanger tube based on SMA wire coils and experimentally assessed the heat transfer enhancement and pressure drop for this self-adaptive system. In a similar way, Aris et al. [85] later investigated the effectiveness of SMA delta wings acting as VGs for convective cooling, using air as working fluid, and demonstrated their impact on heat transfer enhancement. Within liquid cooling, Vilarrubí et al. [86] evaluated the impact of an SMA wing, trained through a Two-Way Shape Memory Effect, on the thermal resistance of a water channel. Their experimental results showcased a reduction in the thermal resistance between 57% and 63%, which was more significant at higher flow rates where the effect of the vortex generators became more noticeable. Additionally, the study highlighted the ability of adaptive SMA wings to maintain a uniform surface temperature even as heat flux increased, obtaining a surface temperature gradient of 7°C for a heat flux variation of 34.2 W/cm^2 (**Figure 8**). Later, Regany et al. [87] experimentally assessed the thermal improvement capacity of an array of SMA adaptive fins inside a liquid

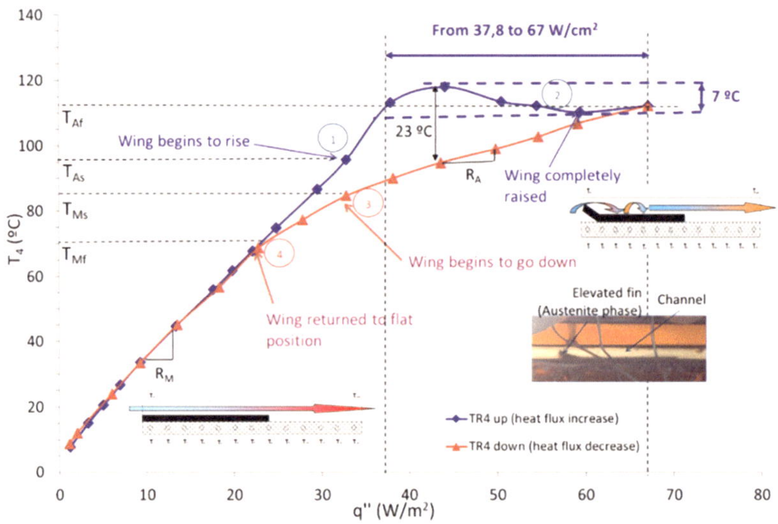

Figure 8.
Surface temperature evolution for varying heat flux due to the adaptive use of an SMA wing [86].

cooling channel, and demonstrated enhancements of 63% in the temperature uniformity and 50% in heat transfer increment due to the implementation of the adaptive fins (**Figure 9**). At the same time, Chu et al. [88] investigated the application of an SMA coil as an adaptive vortex generator to alter the flow characteristics of a micro-channel heat exchanger, allowing to increase the Nusselt number by 112% at a heat flux of 1×10^5 W/m^2. Furthermore, the authors subsequently proposed employing graded SMA coils to intelligently recognize the location of hotspots and experimentally demonstrated better overall performance with this improved system to address random hotspots [89].

Adhering to the same working principle but using a different mechanism, Vilarrubí et al. [90] introduced the use of doubly clamped beams with eccentricity as passive thermal actuators, to act as adaptive vortex generators inside a MCHS. At a certain temperature, these beams would buckle, thereby disrupting the boundary layer of the channel and enhancing its thermal resistance. For a given thermal load variable in time and space, the authors computed a pumping power reduction of 8% in the cooling system (**Figure 10**). Subsequently, the same authors developed a system of self-adaptive passive thermal vortex generators based on bimetallic fins that could adjust their shape with temperature fluctuations (**Figure 11**). This innovative system allows the optimization of the cold plate thermal capacity for any variable heat load

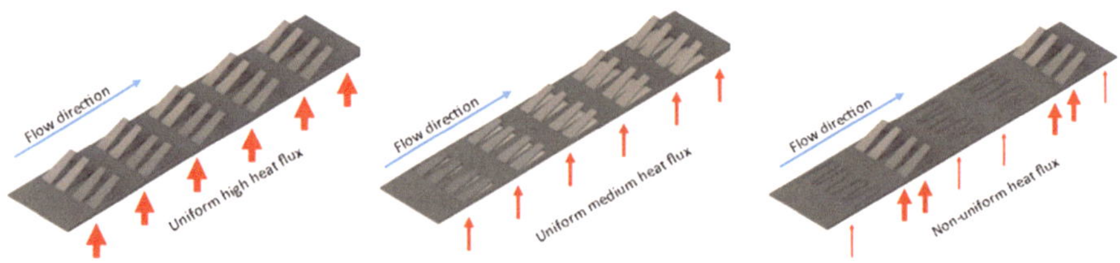

Figure 9.
Working principle of the SMA self-adaptive fins placed inside a liquid cooling channel and acting as vortex generators [87].

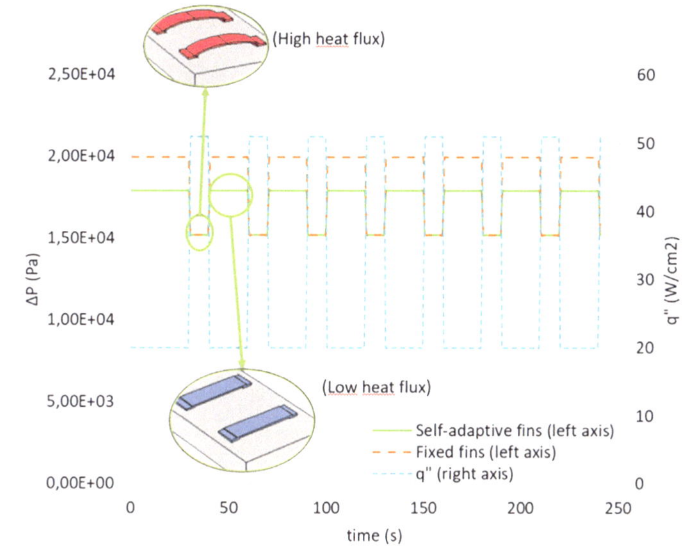

Figure 10.
Comparison of pressure drops along a microchannel with adaptive and fixed doubly clamped beams acting as vortex generators [90].

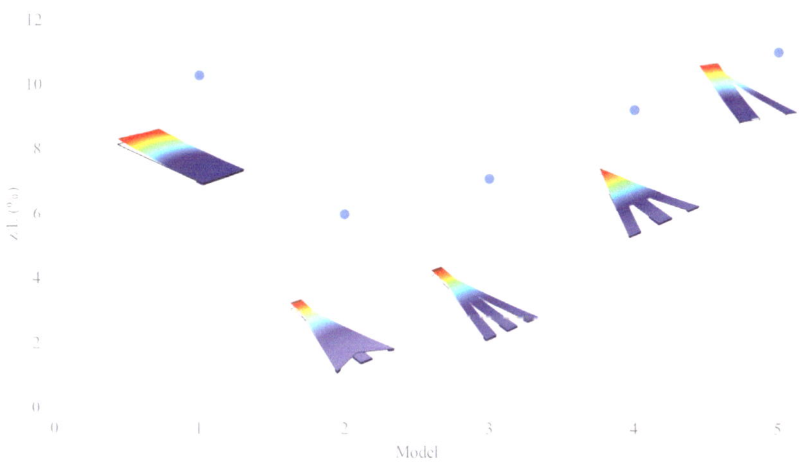

Figure 11.
Vertical displacement (in % of the fin length) reached by different structures of bimetallic adaptive fins [91].

scenario while reducing the energy requirements of the cooling system. The numerical results of this study revealed a heat transfer improvement of up to 40% in a MCHS due to the implementation of the fins and a pumping power reduction of 34% for a specific heat load scenario when compared to a non-adaptive system [91].

Apart from the previously cited works that can adapt their shape based on temperature, there are other research lines involving the use of active actuators to enhance the thermal capacity of cooling systems. For example, Lambert and Pangel [92] proposed employing thin elastic flaps externally actuated to enhance fluid mixing in a microchannel. Their results showed that larger flap displacements resulted in higher mixing fractions, and the addition of multiple flaps further

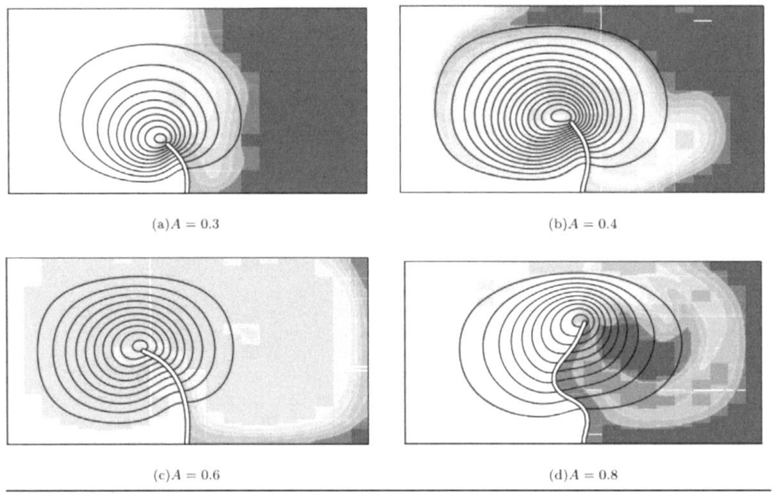

Figure 12.
Streamlines and concentration contours for different aspect ratios of flap length and height of the elastic flap externally actuated [92].

improved mixing (**Figure 12**). Similarly, Mirzaee et al. [93] conducted numerical simulations of a 2D channel with an oscillation-controlled flap, comparing the thermal performances between a flexible and a rigid flap. The findings indicated that the flexible flat not only increased the thermal performance but also led to lower pressure drops compared to the use of rigid flaps. Separately, Fang et al. [94] developed an active heat transfer solution using an array of synthetic jets that generated periodic disturbances when interacting with microchannel flow. The local turbulence created by the jet near the heated channel wall disrupted the thermal boundary layer, resulting in a 130% enhancement in heat transfer with a small increase in pressure drop.

3.3 Other methods

Other methods developed in the literature for efficiently cooling variable heat loads include thermoelectric coolers (TECs), electrowetting, parallel microchannel distributions, two-phase cooling, and the prediction of the thermal map.

TECs have garnered attention as an active cooling method for electronics in recent years. Based on the Peltier effect, TECs can directly transfer electricity to the temperature difference with no moving parts, fast thermal response, silence, and reliability [95]. Although TECs are commonly used with air-cooled heat sinks, some researchers explored their potential in combination with liquid-cooled systems to efficiently manage hotspots. For instance, Hao et al. [96] proposed a hotspot mitigation equipment that integrated TECs with a mini-channel heat sink. Their results showed a decrease in the hotspot temperature of 3.1°C when the R_{th} was 0.6 K/W.

As an alternative approach, some authors evaluated the potential of using the electrowetting phenomenon in microelectronics cooling, employing an array of liquid droplets that can be independently moved along the surface. While originally developed for biological and chemical lab-on-chip applications, this technology has shown its applicability as an adaptive cooling platform for microelectronics. Paik et al. [97] presented an innovative chip cooling technique based on a "digital microfluidic" platform, showcasing the ability to program droplets to cool hotspots in both closed and open systems. Later, other researchers delved into the potential of electrowetting-on-dielectric (EWOD) digital microfluidic devices to create adaptive cooling solutions for non-uniform and transient heat load scenarios [98–100].

Alternatively, Farnam et al. [101] developed a 3D numerical model to assess the thermal behavior of an entire microchannel heat sink coupled with a microprocessor, considering two-way fluid flow under variable heat load scenarios in both time and space. The obtained results revealed that the implementation of a two-way fluid flow could decrease temperature gradients in the device, particularly when hotspots were located downstream in a one-way fluid flow application. These findings highlight the potential for implementing a smart sink capable of adjusting flow direction as needed to control thermal gradients. Furthermore, Ansari and Kim [102] analyzed the thermal performance of double-layer microchannel heat sinks (DL-MCHS) under random hotspots and stated that the cross-channel design of DL-MCHS exhibited the lowest thermal resistance and minimum temperature variation among the hotspots (**Figure 13**). In a different study, Maganti et al. [103] experimentally evaluated the use of various configurations of parallel microchannels coupled with nanofluids to efficiently cool non-uniform heat loads based on different hotspot distributions. Nanofluids were employed as a smart cooling system to efficiently cool hotspots due to the nanoparticle slip mechanisms.

In 2003, Mukherhee and Mudawar [104] demonstrated a smart pumpless loop capable of enhancing its cooling capacity by increasing the velocity of a two-phase mixture along the boiling surface when an increase in the heat flux was detected. Although two-phase cooling for hotspot mitigation has garnered significant interest in recent years, the instabilities associated with this technology have made it challenging to apply on a large scale up to the present time [105].

In a separate approach, Chauhan et al. [106] proposed to address the design of microprocessors to mitigate hotspots, as the placement and arrangement of different electronic components can influence their thermal behavior. The device was cooled using single-phase MCHS using water as the coolant. The study demonstrated that the efficiency of the MCHS was maximized when the higher heat flux components were placed at inlets, hence, an efficient installation of the MCHS proved highly effective in bringing down the maximum hotspot temperature to 72.1°C and limiting the temperature of the remaining areas of the chip to 55°C.

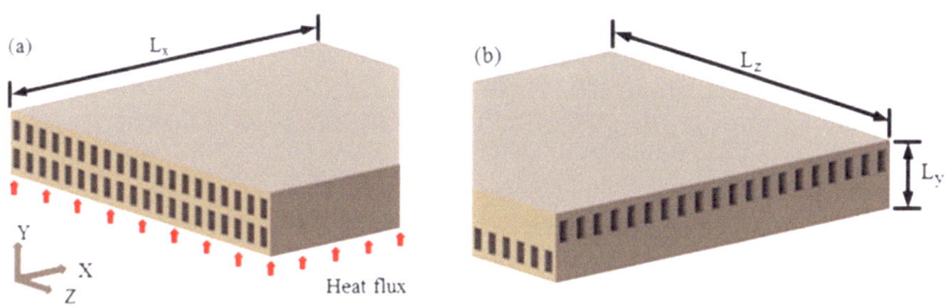

Figure 13.
Schematic diagram of two double-layer microchannel heat sinks in parallel (a) and crossed (b) configurations [102].

4. Conclusions

This chapter summarized different heat transfer enhancement techniques employed to improve the performance of micro heat sinks and overcome their limitations, focusing on intelligent and adaptive solutions that optimize cooling efficiency based on local and instantaneous cooling needs for non-uniform and time-dependent power distribution maps.

Current advanced microelectronic systems exhibit highly non-uniform heat load scenarios, resulting in the appearance of critical hotspot regions that require efficient thermal management. These hotspot regions, typically localized, can dissipate significantly more heat than the rest of the chip, but constant large flow rates across all cooled devices can result in overcooled systems. Moreover, the inability of fixed systems to adapt their behavior to changing heat load scenarios results in oversized pumping powers and poor temperature uniformity for varying operational parameters. To tackle these challenges, advanced cooling solutions that aim to improve the performance of current microelectronics systems must be capable of dynamically adapting to changing boundary conditions across time and space. These solutions should aim for reduced energy consumption and improved thermal performance, with lower thermal resistances and high-temperature uniformity, to obtain efficient and optimized cooling systems under variable conditions.

Recent developments in smart cooling solutions have focused on optimizing the performance of micro heat sinks under varying boundary conditions. Such cooling systems should be able to maximize their thermal performance during peak heat extraction demands while minimizing energy consumption during periods of lower cooling demands. Also, these systems must consider their reliability, since they usually incorporate moving parts that can be more prone to fatigue and wear.

However, the current technology for thermal management of hotspots, especially automatically adaptive cooling, is not fully mature. Thus, there is an urgent need for effective and economical processes to cool random hotspots. In the future, the development of temperature-sensitive materials with better physical performance, lower costs, reliability, and mass production may offer an effective and practical solution for hotspot thermal management.

Acknowledgements

This work is funded by the European Union—NextGenerationEU and Ministry of Universities through the Grants for requalification of the Spanish university system for 2021–2023, Margarita Salas modality.

The author would also like to thank Generalitat de Catalunya for the project awarded to their research group (2021 SGR 01370).

This work is part of the grant PID2021-123634OB-I00, funded by MCIN and FEDER "A way of making Europe".

Conflict of interest

The author declares no conflict of interest.

Author details

Montse Vilarrubí[1,2]

1 Universitat de Lleida, Lleida, Spain

2 Universal Smart Cooling S.L., Spain

*Address all correspondence to: montse.vilarrubi@udl.cat

References

[1] Moore GE. Cramming more components onto integrated circuits. Proceedings of the IEEE. 1998;**86**:82-85

[2] Dennard RH, Cai J, Kumar A. A perspective on today's scaling challenges and possible future directions. Solid State Electronics. 2007;**51**:518-525

[3] Garimella SV, Persoons T, Weibel JA, et al. Electronics thermal management in information and communications technologies: Challenges and future directions. IEEE Transactions on Components, Packaging and Manufacturing Technology. 2017;**7**:1191-1205

[4] Masanet E, Shehabi A, Lei N, et al. Recalibrating global data center energy-use estimates. Science (80-). 2020;**367**:6481

[5] Belkhir L, Elmeligi A. Assessing ICT global emissions footprint: Trends to 2040 & recommendations. Journal of Cleaner Production. 2018;**177**:448-463

[6] Andrews D, Newton EJ, Adibi N, et al. A Circular Economy for the Data Centre Industry: Using Design Methods to Address the Challenge of Whole System Sustainability in a Unique Industrial Sector. Sustainability. 2021;**13**:11

[7] Zhang Y, Li C, Pan M. Design and performance research of integrated indirect liquid cooling system for rack server. International Journal of Thermal Sciences. 2023;**184**

[8] Luo Y, Andresen J, Clarke H, et al. A decision support system for waste heat recovery and energy efficiency improvement in data centres. Applied Energy. 2019;**250**:1217-1224

[9] Roy R, Das S, Labbe B, et al. Co-design of Thermal Management with System Architecture and Power Management for 3D ICs. In: Proceedings Electron Components Technology Conference. San Diego, USA: Institute of Electrical and Electronics Engineers Inc.; 2022. pp. 211-220

[10] Kadam ST, Kumar R. Twenty first century cooling solution: Microchannel heat sinks. International Journal of Thermal Sciences. 2014;**85**:73-92

[11] Tuckerman DB, Pease RFW. High-performance heat sinking for VLSI. IEEE Electron Device Letters. 1981;**2**:126-129

[12] Datta A, Sanyal D, Agrawal A, et al. A review of liquid flow and heat transfer in microchannels with emphasis to electronic cooling. Sadhana—Academy Proceedings in Engineering Sciences. 2019;**44**(12)

[13] He Z, Yan Y, Zhang Z. Thermal management and temperature uniformity enhancement of electronic devices by micro heat sinks: A review. Energy. 2021;**216**:119223

[14] Kandlikar SG, Bapat AV. Evaluation of jet impingement, spray and microchannel chip cooling options for high heat flux removal. Heat Transfer Engineering. 2007;**28**:911-923

[15] Kheirabadi AC, Groulx D. Cooling of server electronics: A design review of existing technology. Applied Thermal Engineering. 2016;**105**:622-638

[16] Zhang X, Jaluria Y. Reliability-based optimization and design limits of microchannel cooling systems. International Journal of Heat and Mass Transfer. 2020;**149**

[17] Gupta M, Bhargava L, Indu S. Mapping techniques in multicore

processors: Current and future trends. The Journal of Supercomputing. 2021;**77**:9308-9363

[18] Sharma CS, Tiwari MK, Poulikakos D. A simplified approach to hotspot alleviation in microprocessors. Applied Thermal Engineering. 2016;**93**:1314-1323

[19] Hamann HF, Weger A, Lacey JA, et al. Hotspot-limited microprocessors: Direct temperature and power distribution measurements. IEEE Journal of Solid-State Circuits. 2007;**42**:56-65

[20] Mccreary CA, Solovitz SA. Development of power law design tool for hotspot mitigation using parallel microchannel heat exchanger. In: 22nd IEEE Intersociety Conference on Thermal and Thermomechanical Phenomena in Electronic Systems—ITHERM, Orlando, USA. Institute of Electrical and Electronics Engineers Inc.; 2023. pp. 0-7

[21] Kearney D, Hilt T, Pham P. A liquid cooling solution for temperature redistribution in 3D IC architectures. Microelectronics Journal. 2012;**43**:602-610

[22] Kandlikar SG. Review and projections of integrated cooling systems for three-dimensional integrated circuits. Journal of Electronic Packaging. 2014;**136**:24001

[23] Waddell AM, Punch J, Stafford J, et al. The characterization of a low-profile channel–confined jet for targeted hot-spot cooling in microfluidic applications. International Journal of Heat and Mass Transfer. 2016;**101**:620-628

[24] Lee YJ, Lee PS, Chou SK. Hotspot mitigating with obliquely finned microchannel heat sink-an experimental study. IEEE Transactions on Components, Packaging and Manufacturing Technology. 2013;**3**:1332-1341

[25] Collin L-M, Colonna J-P, Coudrain P, et al. Hot spot aware microchannel cooling add-on for microelectronic chips in mobile devices. In: 16th IEEE InterSociety Conference on Thermal and Thermomechanical Phenomena in Electronic Systems, Orlando, USA. Institute of Electrical and Electronics Engineers Inc.; 2017. pp. 460-464

[26] Barrau J, Chemisana D, Rosell JI, et al. An experimental study of a new hybrid jet impingement/micro-channel cooling scheme. Applied Thermal Engineering. 2010;**30**:2058-2066

[27] Riera S, Barrau J, Omri M, et al. Stepwise varying width microchannel cooling device for uniform wall temperature: Experimental and numerical study. Applied Thermal Engineering. 2015;**78**:30-38

[28] Vilarrubí M, Riera S, Ibañez M, et al. Experimental and numerical study of micro-pin-fin heat sinks with variable density for increased temperature uniformity. International Journal of Thermal Sciences. 2018;**132**:424-434

[29] Sharma CS, Schlottig G, Brunschwiler T, et al. A novel method of energy efficient hotspot-targeted embedded liquid cooling for electronics: An experimental study. International Journal of Heat and Mass Transfer. 2015;**88**:684-694

[30] Rao X, Jin C, Zhang H, et al. A hybrid microchannel heat sink with ultra-low pressure drop for hotspot thermal management. International Journal of Heat and Mass Transfer. 2023;**211**:124201

[31] Lu G, Yang J, Zhai X, et al. Hotspot thermal management in microchannel heat sinks with vortex generators. International Journal of Thermal Sciences. 2021;**161**:106727

[32] Renfer A, Tiwari MK, Tiwari R, et al. Microvortex-enhanced heat transfer in 3D-integrated liquid cooling of electronic chip stacks. International Journal of Heat and Mass Transfer. 2013;**65**:33-43

[33] Wang J, Yu K, Ye M, et al. Effects of pin fins and vortex generators on thermal performance in a microchannel with Al_2O_3 nanofluids. Energy. 2022;**239**

[34] Tao WQ, He YL, Wang QW, et al. A unified analysis on enhancing single phase convective heat transfer with field synergy principle. International Journal of Heat and Mass Transfer. 2002;**45**:4871-4879

[35] Steinke ME, Kandlikar SG. Single-phase heat transfer enhancement techniques in microchannel and minichannel flows. In: ASME 2nd International Conference on Microchannels and Minichannels, Rochester, NY, USA. American Society of Mechanical Engineers. 2004. pp. 141-148

[36] Sidik NAC, Muhamad MNAW, Japar WMAA, et al. An overview of passive techniques for heat transfer augmentation in microchannel heat sink. International Communications in Heat and Mass Transfer. 2017;**88**:74-83

[37] Dewan A, Srivastava P. A review of heat transfer enhancement through flow disruption in a microchannel. Journal of Thermal Science. 2015;**24**:203-214

[38] Zhou J, Hatami M, Song D, et al. Design of microchannel heat sink with wavy channel and its time-efficient optimization with combined RSM and FVM methods. International Journal of Heat and Mass Transfer. 2016;**103**:715-724

[39] Lin L, Zhao J, Lu G, et al. Heat transfer enhancement in microchannel heat sink by wavy channel with changing wavelength/amplitude. International Journal of Thermal Sciences. 2017;**118**:423-434

[40] Bayrak E, Olcay AB, Serincan MF. Numerical investigation of the effects of geometric structure of microchannel heat sink on flow characteristics and heat transfer performance. International Journal of Thermal Sciences. 2019;**135**:589-600

[41] Ermagan H, Rafee R. Numerical investigation into the thermo-fluid performance of wavy microchannels with superhydrophobic walls. International Journal of Thermal Sciences. 2018;**132**:578-588

[42] Chu JC, Teng JT, Greif R. Experimental and numerical study on the flow characteristics in curved rectangular microchannels. Applied Thermal Engineering. 2010;**30**:1558-1566

[43] Mohammed HA, Gunnasegaran P, Shuaib NH. Numerical simulation of heat transfer enhancement in wavy microchannel heat sink. International Communications in Heat and Mass Transfer. 2011;**38**:63-68

[44] Wang H, Chen Z, Gao J. Influence of geometric parameters on flow and heat transfer performance of micro-channel heat sinks. Applied Thermal Engineering. 2016;**107**:870-879

[45] Ghani IA, Sidik NAC, Mamat R, et al. Heat transfer enhancement in microchannel heat sink using hybrid technique of ribs and secondary channels. International Journal of Heat and Mass Transfer. 2017;**114**:640-655

[46] Shi X, Li S, Mu Y, et al. Geometry parameters optimization for a microchannel heat sink with secondary flow channel. International Communications in Heat and Mass Transfer. 2019;**104**:89-100

[47] Xu JL, Gan YH, Zhang DC, et al. Microscale heat transfer enhancement using thermal boundary layer redeveloping concept. International Journal of Heat and Mass Transfer. 2005;**48**:1662-1674

[48] Zhou F, Zhou W, Qiu Q, et al. Investigation of fluid flow and heat transfer characteristics of parallel flow double-layer microchannel heat exchanger. Applied Thermal Engineering. 2018;**137**:616-631

[49] Srivastava P, Patel RI, Dewan A. A study on thermal characteristics of double-layered microchannel heat sink: Effects of bifurcation and flow configuration. International Journal of Thermal Sciences. 2021;**162**:106791

[50] Debbarma D, Pandey KM, Paul A. Performance enhancement of double-layer microchannel heat sink by employing dimples and protrusions on channel sidewalls. Mathematical Problems in Engineering. 2022

[51] Savino S, Nonino C. Numerical analysis of double-layered microchannel heat sinks with different microchannel heights. Journal of Physics Conference Series. 2020;**1599**

[52] Ghahremannezhad A, Xu H, Alhuyi Nazari M, et al. Effect of porous substrates on thermohydraulic performance enhancement of double layer microchannel heat sinks. International Journal of Heat and Mass Transfer. 2019;**131**:52-63

[53] Bayer Ö, Baghaei Oskouei S, Aradag S. Investigation of double-layered wavy microchannel heatsinks utilizing porous ribs with artificial neural networks. International Communications in Heat and Mass Transfer. 2022;**134**

[54] Gamrat G, Favre-Marinet M, Le Person S. Modelling of roughness effects on heat transfer in thermally fully-developed laminar flows through microchannels. International Journal of Thermal Sciences. 2009;**48**:2203-2214

[55] Lu H, Xu M, Gong L, et al. Effects of surface roughness in microchannel with passive heat transfer enhancement structures. International Journal of Heat and Mass Transfer. 2020;**148**:119070

[56] Singh S, Singh SK, Mali HS, et al. Numerical investigation of heat transfer in structured rough microchannels subjected to pulsed flow. Applied Thermal Engineering. 2021;**197**:117361

[57] Mandev E, Manay E. Effects of surface roughness in multiple microchannels on mixed convective heat transfer. Applied Thermal Engineering. 2022;**217**:119102

[58] Sajid MU, Ali HM. Recent advances in application of nanofluids in heat transfer devices: A critical review. Renewable and Sustainable Energy Reviews. 2019;**103**:556-592

[59] Maghrabie HM, Olabi AG, Sayed ET, et al. Microchannel heat sinks with nanofluids for cooling electronic components: Performance enhancement, challenges, and limitations. Thermal Science and Engineering Progress. 2023;**37**:101608

[60] Ahmed HE, Mohammed HA, Yusoff MZ. An overview on heat transfer augmentation using vortex generators and nanofluids: Approaches and

applications. Renewable and Sustainable Energy Reviews. 2012;**16**:5951-5993

[61] Heydari A, Akbari OA, Safaei MR, et al. The effect of attack angle of triangular ribs on heat transfer of nanofluids in a microchannel. Journal of Thermal Analysis and Calorimetry. 2018;**131**:2893-2912

[62] Khan MZU, Akbar B, Sajjad R, et al. Investigation of heat transfer in dimple-protrusion micro-channel heat sinks using copper oxide nano-additives. Case Studies in Thermal Engineering. 2021;**28**

[63] Aguirre I, González A, Castillo E. Numerical study on the use of shear-thinning nanofluids in a micro pin-fin heat sink including vortex generators and changes in pin shapes. Journal of the Taiwan Institute of Chemical Engineers. 2022;**136**

[64] Mohammadpour J, Salehi F, Lee A, et al. Nanofluid heat transfer in a microchannel heat sink with multiple synthetic jets and protrusions. International Journal of Thermal Sciences. 2022;**179**:107642

[65] Alkasmoul FS, Asaker M, Almogbel A, et al. Combined effect of thermal and hydraulic performance of different nanofluids on their cooling efficiency in microchannel heat sink. Case Studies of Thermal Engineering. 2022;**30**:101776

[66] Biswas G, Chattopadhyay H, Sinha A. Augmentation of heat transfer by creation of streamwise longitudinal vortices using vortex generators. Heat Transfer Engineering. 2012;**33**:406-424

[67] Chai L, Xia GD, Wang HS. Numerical study of laminar flow and heat transfer in microchannel heat sink with offset ribs on sidewalls. Applied Thermal Engineering. 2016;**92**:32-41

[68] Ebrahimi A, Kheradmand S. Numerical simulation of performance augmentation in a plate fin heat exchanger using winglet type vortex generators. International Journal of Mechanical Engineering and Mechatronics. 2012

[69] Chai L, Xia GD, Wang HS. Laminar flow and heat transfer characteristics of interrupted microchannel heat sink with ribs in the transverse microchambers. International Journal of Thermal Sciences. 2016;**110**:1-11

[70] Mohammadi A, Koşar A. Review on heat and fluid flow in micro pin fin heat sinks under single-phase and two-phase flow conditions. Nanoscale and Microscale Thermophysical Engineering. 2018;**22**:153-197

[71] Lan J, Xie Y, Zhang D. Flow and heat transfer in microchannels with dimples and protrusions. Journal of Heat Transfer. 2012;**134**

[72] Azarkish H, Barrau J, Coudrain P, et al. Self-adaptive microvalve array for energy efficient fluidic cooling in microelectronic systems. In: 16th IEEE InterSociety Conference on Thermal and Thermomechanical Phenomena in Electronic Systems, Orlando, FL, USA. IEEE; 2017. pp. 522-529

[73] Amnache A, Laguna G, Léveillé É, et al. Fabrication and demonstration of a self-adaptive microvalve array for distributed liquid cooling in microelectronic interposers. Journal of Microelectromechanical Systems. 2020;**29**:769-775

[74] Laguna G, Ibanez M, Rosell J, et al. Experimental validation of a smart microfluidic cell cooling solution. In: InterSociety Conference on Thermal and Thermomechanical Phenomena in Electronic Systems, ITHERM. Virtual: IEEE; 2020

[75] Waddell AM, Punch J, Stafford J, et al. The heat transfer performance in a square channel downstream of a representative shape memory alloy structure for microfluidics applications. In: Annual IEEE Semiconductor Thermal Measurement and Management Symposium, San Jose, USA. IEEE; 2015. pp. 273-279

[76] Vilarrubí M, Beberide D, Regany D, et al. Self-adaptive flow regulation through shape memory alloy valves for microelectronics cooling. In: IEEE, editor. 2023 22nd IEEE Intersociety Conference on Thermal and Thermomechanical Phenomena in Electronic Systems (ITherm). Orlando, FL: IEEE Computer Society; 2023. pp. 1-8

[77] Tudor A, Saez J, Florea L, et al. Poly(ionic liquid) thermo-responsive hydrogel microfluidic actuators. Sensors & Actuators, B: Chemical. 2017;**247**:749-755

[78] Yan Y, Zhang C, Liu Y, et al. Numerical study on hotspots adaptive cooling and thermal-hydraulic performance enhancement of fractal microchannel heat sink embedded with hydrogels. International Journal of Thermal Sciences. 2022;**172**:107272

[79] Li X, Xuan Y. Embedded cooling of high-heat-flux hotspots using self-adaptive microchannel/pin-fin hybrid heat sink. IEEE Transactions on Components, Packaging and Manufacturing Technology. 2022;**12**:1627-1635

[80] Li X, Xuan Y. Self-adaptive cooling of chips with unevenly distributed high heat fluxes. Applied Thermal Engineering. 2022;**202**:117913

[81] Zhang TJ, Wen JT, Julius A, et al. Extremum seeking micro-thermal-fluid control for active two-phase microelectronics cooling. In: Proceedings of the IEEE Conference on Decision and Control. 2010;**2010**:1899-1904

[82] Da Luz S, Kattinger G, Laguna G, et al. Variable pumping control for low power microfluidic chip cooling. In: 24th International Workshop on Thermal Investigations of ICs and Systems, Proceedings (THERMINIC), Stockholm, Sweden. IEEE; 2018. pp. 1-5

[83] Shahi P, Deshmukh AP, Hurnekar HY, et al. Design, development, and characterization of a flow control device for dynamic cooling of liquid-cooled servers. Journal of Electronic Packaging, Transactions of the ASME. 2021;**144**:1-9

[84] Champagne PR, Bergles AE. Development and testing of a novel, variable-roughness technique to enhance, on demand, heat transfer in a single-phase heat exchanger. Journal of Enhanced Heat Transfer. 2001;**8**:341-352

[85] Aris MS, Owen I, Sutcliffe CJ. The development of active vortex generators from shape memory alloys for the convective cooling of heated surfaces. International Journal of Heat and Mass Transfer. 2011;**54**:3566-3574

[86] Vilarrubí M, Morell G, Rosell J, et al. Experimental characterization of a self-adaptive shape memory alloy cooling approach to regulate temperature under varying heat loads. International Journal of Heat and Mass Transfer. 2019;**139**:632-640

[87] Regany D, Majós F, Barrau J, et al. Design and test of shape memory alloy fins for self-adaptive liquid cooling device. Applied Thermal Engineering. 2022;**206**:118010

[88] Chu X, You H, Tang X, et al. Smart microchannel heat exchanger based on

the adaptive deformation effect of shape memory alloys. Energy Conversion and Management. 2021;**250**:114910

[89] Chu X, Tang X, You H, et al. Smart microchannel heat exchangers with built-in graded SMA vortex generators that responds to random hotspots. Applied Thermal Engineering. 2023;**226**:120261

[90] Vilarrubi M, Laguna G, Amnache A, et al. Thermostatic fins for spatially and temporally adaptive microfluidic cooling. In: 24th International Workshop on Thermal Investigations of ICs and Systems, Proceedings (THERMINIC), Stockholm, Sweden. IEEE; 2018

[91] Vilarrubi M, Regany D, Majos FX, et al. Numerical evaluation of bimetallic self-adaptive fins acting as flow disturbing elements inside a microchannel. In: InterSociety Conference on Thermal and Thermomechanical Phenomena in Electronic Systems, ITHERM. San Diego: IEEE; 2022. pp. 1-7

[92] Lambert RA, Rangel RH. The role of elastic flap deformation on fluid mixing in a microchannel. Physics of Fluids. 2010;**22**:1-15

[93] Mirzaee H, Dadvand A, Mirzaee I, et al. Heat transfer enhancement in microchannels using an elastic vortex generator. Journal of Enhanced Heat Transfer. 2012;**19**:199-211

[94] Fang R, Khan JA. Active heat transfer enhancement in single-phase microchannels by using synthetic jets. Journal of Thermal Science and Engineering Applications. 2013;**5**:011006

[95] Chen WY, Shi XL, Zou J, et al. Thermoelectric coolers for on-chip thermal management: Materials, design, and optimization. Materials Science & Engineering R: Reports. 2022;**151**:100700

[96] Hao X, Peng B, Xie G, et al. Efficient on-chip hotspot removal combined solution of thermoelectric cooler and mini-channel heat sink. Applied Thermal Engineering. 2016;**100**:170-178

[97] Paik PY, Pamula VK, Chakrabarty K. Adaptive cooling of integrated circuits using digital microfluidics. IEEE Transactions on Very Large Scale Integration (VLSI) Systems. 2008;**16**:432-443

[98] Cheng JT, Chen CL. Active thermal management of on-chip hot spots using EWOD-driven droplet microfluidics. Experiments in Fluids. 2010;**49**:1349-1357

[99] Bindiganavale GS, Amaya M, Moon H. Demonstration of hotspot cooling using digital microfluidic device. Journal of Micromechanics and Microengineering. 2018;**28**

[100] Ahmad I, Ranjan A, Pathak M, et al. A wettability-mediated microdroplet under electrowetting effect for hotspot cooling. IEEE Transactions on Components, Packaging and Manufacturing Technology. 2022;**12**:288-296

[101] Farnam D, Sammakia B, Ghose K. Development of a complete transient microchannel heat sink model. In: 11th IEEE InterSociety Conference on Thermal and Thermomechanical Phenomena in Electronic Systems, Orlando, FL, USA. IEEE; 2008. pp. 113-120

[102] Ansari D, Kim KY. Performance analysis of double-layer microchannel heat sinks under non-uniform heating

conditions with random hotspots. Micromachines. 2017;**8**

[103] Maganti LS, Dhar P, Sundararajan T, et al. Mitigating non-uniform heat generation induced hot spot(s) in multicore processors using nanofluids in parallel microchannels. International Journal of Thermal Sciences. 2018;**125**:185-196

[104] Mukherjee S, Mudawar I. Smart pumpless loop for micro-channel electronic cooling using flat and enhanced surfaces. IEEE Transactions on Components, Packaging and Manufacturing Technology. 2003;**26**:99-109

[105] Huang Y, Yang Q, Miao J, et al. Experimental investigation on flow boiling characteristics of a radial micro pin–fin heat sink for hotspot heat dissipation. Applied Thermal Engineering. 2023;**219**:119622

[106] Chauhan A, Sammakia B, Ghose K, et al. Hot spot mitigation using single-phase microchannel cooling for microprocessors. In: 12th IEEE InterSociety Conference on Thermal and Thermomechanical Phenomena in Electronic Systems (ITHERM), Las Vegas, NV, USA. IEEE; 2010. pp. 1-11